Desalination - Ecological Consequences

Edited by Karthick Ramalingam and Akif Zeb

Published in London, United Kingdom

Desalination - Ecological Consequences
http://dx.doi.org/10.5772/intechopen.104053
Edited by Karthick Ramalingam and Akif Zeb

Contributors
Alexander Fayer, Yunusa Muhammad Ishiaku, Usman Abdullahi, Haruna Ibrahim, Muhammed Rabiu Hassan, Safiyanu Abdullahi Ahmed, Dejen Tsegaye, Wafa Suwaileh, Rima Isaifan, Mohammad Ahmed, Reza Rahighi, Amirmahmoud Bakhshayesh, Adel Hussein Abouzied, Mehdi Sepahvand, Yehia Aly A. Shebl

First published in London, United Kingdom, 2023 by IntechOpen
IntechOpen is the global imprint of INTECHOPEN LIMITED, registered in England and Wales, registration number: 11086078, 5 Princes Gate Court, London, SW7 2QJ, United Kingdom

British Library Cataloguing-in-Publication Data
A catalogue record for this book is available from the British Library

Additional hard and PDF copies can be obtained from orders@intechopen.com

Desalination - Ecological Consequences
Edited by Karthick Ramalingam and Akif Zeb
p. cm.
Print ISBN 978-1-83768-673-5
Online ISBN 978-1-83768-674-2
eBook (PDF) ISBN 978-1-83768-675-9

For EU product safety concerns:
IN TECH d.o.o., Prolaz Marije Krucifikse Kozulić 3, 51000 Rijeka, Croatia,
info@intechopen.com or visit our website at intechopen.com.

Meet the editors

Dr. Karthick Ramalingam is an electrochemist and photoelectrochemist. He completed his doctorate program at the Central Electrochemical Research Institute (CSIR), India, in 2018. He was a postdoctoral fellow (2018–2020) and research professor (2020–2022) at South China Normal University. He has published twenty-one articles in peer-reviewed journals and attended many international conferences. He received funding from the Foreign Young Talent Program of the Ministry of Science and Technology, PR China, in 2021–2022. His work is focused on solar–water–energy nexus technology. Currently, he is an assistant professor at the University of Warsaw, Poland, working in the field of photoanodes development for the photoelectrochemical water splitting process.

Dr. Akif Zeb is a materials scientist and inorganic chemist. He has a Ph.D. in Inorganic Chemistry from the University of Science and Technology of China (USTC), as well as master's degrees in Chemistry and Materials Engineering. In addition to his extensive experience in academia and industry, Dr. Zeb has published several articles in peer-reviewed journals and has presented his work at international conferences. He has won several fellowships and awards, including the Anhui Government Scholarship, Guangdong Talent Fellow, Hunan Talent Fellow, and a Marie Skłodowska-Curie Actions (MSCA) fellowship. His work is focused on developing advanced materials for energy, environment, and catalysis. He is also a blogger and writes on science, society, and education.

Contents

Preface

Desalination – Ecological Consequences delivers a comprehensive and robust exploration of desalination, including theories, experiments, and calculations.

The book is organized into two sections. Section 1 on the economic impacts of desalination includes techno-economic analyses of reverse osmosis and other practices of desalinated water. Section 2 discusses desalination technologies, including advances in harvesting freshwater, as well as the biological impacts of desalination technology.

The book also includes a general discussion about desalination, technologies, materials processing agents, techno-economic analyses of reverse osmosis plants, biological property, household practices, freshwater management and impact in industrial water treatment plants.

Karthick Ramalingam
Laboratory of Molecular Research for Solar Energy Innovations,
Center of New Technologies,
University of Warsaw,
Warszawa, Poland

Akif Zeb
South China Normal University,
Guangzhou, China

Section 1

Economic Impacts of Desalination

Chapter 1

Perspective Chapter: Technical and Economic Analysis of Reverse Osmosis Desalination System

Mehdi Sepahvand

Abstract

Economic thermodynamic analysis is a branch of engineering science that is derived from economic laws. The goal of economic thermodynamic analysis of systems is the lowest price. Price calculation in a system includes the following steps: Determining the actual price of products. Provide a reasonable way to price products. Providing information on which calculations are made. The overall investment cost of a project includes fixed investment costs, including the costs related to the purchase of land, the construction of the necessary facilities and equipment, and the purchase and installation of machinery, as well as the initial costs related to the investment, including a series of other side costs. It is possible that their relationships and the percentage of their allocated costs in the project are explained separately and finally the estimation equations of each part of the power plant cycle as well as the economic modeling of the RO system and the effective input parameters such as the input salt concentration, discharge Feeding and input water, ambient pressure, number and type of membrane, etc. are stated along with their relationships. Finally, a RO system design flowchart and how to solve its algorithm are explained in detail.

Keywords: desalination, reverse osmosis, economic, cost, power plant, MED system

1. Introduction

The need for water all over the world has increased due to the growth of the population and also due to the growth of the industry, and the water resources are rapidly being depleted. Since 1990, more than 80 countries are facing the problem of water shortage, while more than 70% of the earth's surface is covered with water; But only one percent of these resources are suitable for use, and 97.5% of them are oceans. The only solution; The use of salt water desalination techniques can solve the problems of water shortage. The two main types of desalination techniques that are widely used are evaporation methods and membrane methods. Evaporation methods such as MSF and MED are common in regions such as the Middle East that have huge energy resources. In 2007, all over the world, 66% of sweetening was done with the MSF process, and only 22% was using reverse osmosis (RO) [1]. In 2011, this statistic changed to 60% for reverse osmosis and only 27% for MSF, which shows the increasing importance of the reverse osmosis process [2]. Reverse osmosis is a separation process whose driving force is pressure, in which salt water is purified by pressure by passing through a semi-permeable membrane. This process depends on the resistance of the membrane and the concentration of water impurities. Reverse osmosis is a process that consumes a lot of energy. The costs of a reverse osmosis system, which includes investment and operating costs, are classified in the diagram "**Figure 1**". Due to the operation of reverse osmosis membranes at high pressure, the electric energy of feed pumps is an important part of the operating cost of these systems. However, the pressure drop in reverse osmosis systems is low and the concentrated water flow leaves the last membrane with a pressure equal to 80 to 90% of the supply pressure. If the concentrated water of the system is directed to the surface water, this excess pressure must be dissipated before discharge. The pressure that is lost in the flow of concentrated water through the control valve is wasted energy, because it does not do any useful work in the purification system. Due to the high level of pressure and flow rate of condensed water, the amount of wasted energy is significant.

2. Estimation of total capital investment (TCI)

In order to estimate the total investment cost of a project, the costs related to the purchase of land, the construction of necessary facilities and equipment, and the purchase and installation of special machinery and equipment that are used to make the system work should be calculated. These costs are Fixed Capital investment (FCI); But the initial costs related to the investment of a project, in addition to these fixed costs, also include a series of other side costs, the sum of these costs and fixed costs is called total investment costs (TCI).

2.1 Cost estimate of purchased equipment (PEC)

Estimating the cost of purchasing equipment is the first and most important step in estimating the costs of a project. It is clear that the accuracy of estimating these costs

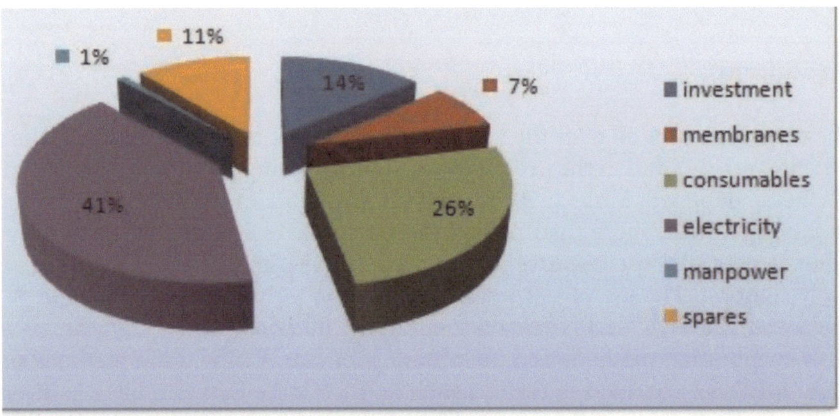

Figure 1.
Classification of seawater reverse osmosis costs for fresh water production (produced water flow rate 125 m³/h, system recovery 40%, one pass and with a life of 10 years) [2].

depends on the amount of information available to engineers. Of course, in addition to the amount of available information, the time and the budget given to the relevant engineers for price estimation will also affect the accuracy of cost estimation. The best way to estimate the cost of purchasing equipment is to refer to the sellers of these items and use their information in this regard. It should be noted that the costs related to the transportation and installation of the equipment should be added separately to the cost of purchasing the equipment. Another way to estimate the cost of purchasing equipment is to refer to previous purchases and use the information in them. In such a case, by referring to professional and experienced people whose job is to estimate the price, or by using the information that engineering companies often make available to users, the cost of purchasing equipment can be estimated. In addition to these methods, software packages designed for this purpose can also be used, although it should be noted that the costs estimated by these softwares may be higher than which are calculated through different charts, do not have more accuracy and advantage.

2.1.1 Use of price estimation charts

In cases where referring to equipment sellers is not very useful or the time and budget required for price estimation is insignificant, in such cases, by referring to various brochures that are mostly presented in the form of price estimation charts. Can be estimated the cost of purchasing equipment. Such charts are obtained experimentally and are provided to users by different manufacturers.

2.1.2 The effect of the size of parts on the price of equipment

In all price estimation charts, the purchase cost of equipment is shown in a logarithmic chart according to their size changes. The lines drawn in such diagrams have a slope of α. The value of α plays a very important role in estimating the cost of purchasing equipment; Therefore, you should be careful in choosing it. The relationship that relates the cost of purchasing equipment to α is [3];

$$PEC_y = PEC_w \left(\frac{X_y}{X_w}\right)^{\alpha} \tag{1}$$

By using this relationship, the cost of purchasing equipment (PEC_y) for a desired capacity or size (X_y) can be calculated by having the cost of purchasing equipment (PEC_w) for a known capacity or size (X_w) achieved.

For processes that deal with heat, α is usually smaller than 1; This means that the percentage increase (or decrease) in the purchase cost of equipment is less than the percentage increase (or decrease) in their size or capacity. If there is no information about the desired design, $\alpha = 0.6$ is used. This work is known as the sixteenth law.

2.1.3 Cost indices

All the prices that are examined in the economic analysis must be stated relative to the same year in which those prices were estimated; That is, if we want to use the data related to a specific year for the present, we must also consider the price index and the inflation rate. For this purpose, the following relationship can be used [3];

$$\text{Current time in price equipment} = \text{The price of the equipment in the desired year} \times (\text{Price index for the present})/(\text{Price index for the desired year}) \quad (2)$$

2.2 Purchased equipment installation

The cost of installing the equipment actually includes the cost of transporting the goods from the factory, workers' wages, the cost of emptying the cargo at the place of installation of the equipment, the insurance of the workers and related goods, the foundation of the intended place for the installation of the equipment, and in general, it includes all the costs that have been purchased for the installation of the equipment. They find communication [3].

2.2.1 Piping cost

The cost related to piping includes the cost of the pipe used and also the wages of the workers during the period when the piping of the system is completed [3].

2.2.2 The cost of instrumentation and control

The multiplier value that is considered for these costs depends on the degree of automation of the devices. The more advanced the regulating and controlling devices are, the higher the cost of using them will certainly be. Of course, in cases where the use of advanced computers and complex control systems is more common, this coefficient will have a higher value [3].

2.2.3 The cost of electrical equipment and materials

These costs include the cost of parts used in power distribution lines, current replacement levers, control centers, emergency power stations, etc. [3].

2.2.4 The cost of purchasing or renting land

The cost of purchasing or renting land is significantly dependent on the geographical location of the place in question, and unlike other costs that have been studied before, this cost is likely to increase over time. But it never decreases [3].

2.2.5 The cost of civil, structural and architectural work

This category of costs includes the general costs of construction as well as other services such as the construction of streets, sidewalks and fences in the desired location and the development of green spaces. As seen in **Table 1**, the costs related to this part are variable depending on whether the construction is related to the construction of a new system inside the site, or a new unit inside the site, or the development of a site.

2.3 Costs related to auxiliary equipment

The costs related to auxiliary equipment include all the costs that must be spent so that the main equipment can perform optimally. These costs are often spent on fuel,

Type of Activity	Building a new system inside a new site	Building a new unit on a site that has already been built	Development of an already established site
A process that deals only with solid materials	83%	40%	30%
A process that deals with both solid and liquid materials	62%	44%	22%
A process that only deals with fluids materials	60%	20–33%	21%

Table 1.
Costs related to civil, construction and architectural affairs as a percentage of the cost of purchasing equipment [3].

water, steam, electricity, cooling and sewage management. Eliminating garbage, controlling environmental pollution, providing firefighting equipment, first aid, and building dining halls are among the other uses of these expenses [3].

2.4 Costs related to engineering and supervision and supervision

This category of costs includes costs such as development and construction, preparation of appropriate maps and plans of the desired location, and other costs that are related to engineering matters; Such as the cost of purchasing engineering equipment, the cost of supervision and supervision, the implementation of construction plans, and the wages of consulting engineers [3].

For a better understanding of the issue, the costs and the ratio of each percentage to the cost of purchasing equipment are shown separately in **Tables 1** and **2**.

2.5 The cost of constructing a building including the contractor's wages

These costs include the cost of all temporary equipment and facilities. Among the examples of these equipments, we can mention the living place of the workers, which is temporarily built inside the site, the insurance fee and the wages of the construction workers. It should be noted that these costs are in addition to the construction costs that were mentioned in the previous sections. It should be noted that in the costs related to this part, the contractor's profit and wages are also calculated. The experimental estimate of these costs is equivalent to 15% of direct cost (DC) [3].

2.6 The cost of possible accidents

Sometimes, unpredictable events such as weather changes, sudden stoppage of work, sudden changes in market prices and problems caused by the transportation of goods may have some effect on the estimated costs. For this purpose, a cost is usually included as a cost caused by possible incidents [3].

2.7 Startup costs

The Startup Costs (SUC) of system includes workers' wages, the cost of materials and equipment, and other additional costs that must be spent during the setting up of

Type of Activity	% of the purchase of equipment (PEC)	Different conditions
cost of installing parts	Normally, 20–90%	If there is not much information about the type and amount of equipment, the cost of installing parts can be considered equal to 45% of the cost of purchasing equipment
Piping Cost	10–70%	1. where more solid materials are used and there is no need for piping (10–20%) of the cost of purchasing equipment 2. for systems that mostly work with fluids is 50–70% of the cost of purchasing equipment 3. power plants that are provided with coal fuel, the cost of piping is equal to 16% of the cost of purchasing equipment 4. for systems that work with both solids and fluids, the cost of piping is equal to 31% of the cost of purchasing equipment 5. and for systems that only work with Fluids work; Like Heat Recovery Steam Generators (HRSGs), the cost of piping is estimated to be equal to 66% of the equipment purchase cost
The cost of Instrumentation and control	often allocate 2–30%	1. In general and taking into account expensive and advanced control systems, the common range for the cost of setting and controlling the system is equivalent to 6–40% of the cost of purchasing parts 2. For old steam power plants that use traditional control systems, the amount of this coefficient is 6–10% of the equipment purchase cost 3. In the absence of appropriate information to select the price coefficient of system controls and adjustments, the average and common value of 20% of the purchase cost of parts can be used
The cost of Electrical equipment and materials	approximately 10–15%	1. The average and common amount used for these costs is equivalent to 11% of the equipment purchase cost 2. in some cases where electronic equipment plays a major role in the design; Like power plants, these costs reach up to 48% of the equipment purchase cost
The cost of purchasing or renting land	10%	—
Costs related to auxiliary equipment	30–100%	The average amount that is often considered for these costs is 65% of the equipment purchase cost
Costs related to engineering and supervision and supervision	25–75%	The common amount that is usually considered for these costs is 30% of the equipment purchase cost

Table 2.
Costs related to a percentage of the cost of purchasing equipment [3].

the system. Of course, to these costs, the costs resulting from the decrease in income due to the system being shut down or its operation under partial load should also be added [3].

2.8 Working capital cost

The Working capital (WC) cost of the system depends on the average period of time required to produce the product and reach the customer; The meaning of products reaching the customer is when money is received from the customer for the sale of the products [3].

For a better understanding of the issue, the costs and the ratio of each percentage to the fixed investment cost (FCI) and the total investment cost (TCI) are shown separately in **Table 3**.

2.9 The cost of obtaining a licensing, research and development department

If there is a desire to use franchise to obtain a work permit, in this case, the cost of obtaining a licensing, research and development (LRD) department are directly dependent on the process that is carried out inside the system. In fact, for any type of industrial activity and according to the extent of the activity, these costs have a specific range, so we should add these amounts to the total investment cost. Therefore, there is no standard or conventional amount for these costs.

2.9.1 The cost due to the lack of budget estimated during the construction

The time it takes for a project from the initial design stage to be put into operation and its equipment is launched is between 1 and 5 years. During this period, some of the investment costs should be spent on providing the salaries of the system design engineers and civil engineers, as well as the purchase and installation of system equipment and things like this. In the same way, a large amount of initial capital is spent without obtaining any income. This money may be withdrawn from the company's fund, or taken as a loan from a bank or institution, or a combination of these two cases. In any case, some of the money intended for investment will be spent on things that will not bring any profit or income to the company. In fact, the costs of this section are applied due to the change in the value of money during the construction period; That is, the longer the project takes, the higher the costs of this part will be.

Type of Activity	% of Fixed Capital Investment (FCI)	Different conditions
The cost of possible accidents	5–20%	—
Startup Costs	1–5%	If there is not enough information about the Startup Costs, the conventional value of 10% of the Fixed Capital investment (FCI) is used for the Startup Costs
Working capital cost	% of the Total Investment Cost (TCI). 10–20%	Usually, a conventional value is used to calculate the Working capital cost of the system. This amount is equivalent to 15% of the total investment cost

Table 3.
Costs related to a percentage of the cost of fixed capital investment (FCI) and the Total investment cost (TCI) [3].

3. Simplified relationships related to the initial investment of the project

The purpose of stating the contents mentioned so far is to provide simple methods for estimating the initial investment cost of a new plan or expansion of old plans. After examining the various factors that are effective in estimating the cost of a reverse osmosis system and a multi-stage distillation system, in this part, by creating a mathematical relationship between these factors, we will arrive at simple relationships to estimate the cost of these systems.

As seen, total capital investment (TCI), Fixed Capital investment (FCI), Startup Costs (SUC), Working capital cost (WC), licensing, research and development department cost (LRD) and the estimated cost of underfunding during construction (AFUDC) also direct costs (DC) of the project are calculated, which includes onsite cost (ONSC) and offsite cost (OFSC) [4].

$$TCI = FCI + SUC + WC + LRD + AFUDC \tag{3}$$

$$DC = ONSC + OFSC \tag{4}$$

$$OFSC = \begin{cases} 1.2\ ONSC & \text{new system} \\ 0.45\ ONSC & \text{expansion} \end{cases} \tag{5}$$

$$WC = 0.15TCI \tag{6}$$

$$SUC = 0.1FCI \tag{7}$$

If it is assumed:

$$LRD = AFUDC + 0.15FCI \tag{8}$$

In this case, by combining relations (3),(8) and (6), (7) the following relation is obtained:

$$TCI = 1.47FCI \tag{9}$$

It follows from relations (4), (5), (9):

$$TCI = 1.84DC = 1.84(ONSC + OFSC) \tag{10}$$

By combining relations (6) and (10), the following relation is obtained:

$$TCI = \begin{cases} 4.05\ ONSC & \text{new system} \\ 2.67\ ONSC & \text{expansion} \end{cases} \tag{11}$$

Experience has shown that the cost of fixed investment in a new system is between 2.8 and 5.5 times the cost of purchasing equipment [4]. Therefore:

$$FCI = \begin{cases} 2.8 \sim 5.5\ PEC & \text{new system} \\ 2.83\ PEC & \text{expansion} \end{cases} \tag{12}$$

By combining the above relations and relations (12), the following general relation is obtained:

$$TCI = \begin{cases} 4.12 \sim 8.09\ PEC & \text{new system} \\ 4.16\ PEC & \text{expansion} \end{cases} \tag{13}$$

As it is clear from relations (12) and (13), by having equipment purchase cost (PEC) and internal site costs (ONSC), the total investment cost (TCI) can be estimated. Therefore, you should be as accurate as possible in estimating the cost of purchasing equipment and the internal costs of the site; Because the more accurately the costs are estimated, the more accurate the overall investment cost will be.

Several methods to express the cost of purchasing equipment in terms of design parameters are stated in the equation, but here by using the cost functions for multi-stage distillation and reverse osmosis and other components extracted from references [3–5] respectively has been used.

Z_k is the purchase cost for the k-th component, N is the number of operating hours per year, φ is the maintenance factor, if there is no comprehensive information, the value of 1.06 can be used, and the investment return factor (CRF) which depends on the r_n percentage of inflation and the estimated life It is the equipment that is determined from the following relationship:

$$Z_K = \frac{TCI \times CRF \times \varphi}{N \times 3600} \tag{14}$$

$$CRF = \frac{r_n(1+r_n)^{year}}{(1+r_n)^{year}-1} \tag{15}$$

where year is the useful life of the power plant. With the purchase cost of equipment (Z_k) and the internal costs of the site, the total investment cost can be estimated. The operating cost of multi-stage distillation and reverse osmosis system will be explained in the form of the relationships presented below.

4. Economic modeling of RO and MED system

With the equipment purchase cost (CC) and internal site costs (ONSC), the total investment cost (TCI) can be estimated. Equations for estimating the price of each component of the reverse osmosis and MED system are shown in **Tables 3** and **4**, respectively, and the cost of other available equipment such as steam turbine and condenser are described below. It is worth mentioning that the fixed parameters of the economic analysis are shown in **Tables 5** and **6**.

parameter
Capacity factor MED + RO (fc) [7]
The price of each membrane (Cm[$])
Electricity price (Ce [$/(kW h) − 1])
Inflation percentage [7]
System operation time [7]
Price of pressure vessel($) [7]

Table 4.
To estimate the cost of purchasing equipment, the functions related to the price estimation of the mentioned system can be extracted from ***Table*** *5.*

Equation	Component
$CC_{SWIP} = 996\,(Q_f)^{0.8}$	capital cost of the seawater intake and pre-treatment
$CC_{hpp} = 52\left(Q_{fhpp}\Delta P_f\right)$	capital cost of high pressure pump and pre-treatment
$Z_{Pump} = a_1(mV\Delta P)^{a_2} f_m.\phi_\eta$ $\phi_\eta = 1 + \left(\frac{1-\bar{\eta}_1}{1-\eta_1}\right)^{a_3}$ $f_m = \begin{cases} \text{Cast iron} = 1 \\ \text{Steel} = 1.41 \end{cases}$ That: f_m: correction factor of pipe material, in this case: f_m =1.41 ϕ_η: first law efficiency correctionfactor $a_1 = 549.13\frac{\$}{Kw^{0.71}}$ ، $a_2 = 0.71$ ، $a_3 = 3$ ، $\bar{\eta}_1 = 0.8$	Pump
$CC_m = \sum_{N_{RO}} C_k n_{m,j} n_{PV,j} + \sum_{N_{RO}} C_{PV} n_{PV,j}$	Total membrane module cost
$CC_{px} = 313.47\,Q_{px\ hin}{}^{0.58}$	Pressure exchanger

Table 5.
Price of RO cycle components [6, 7].

Equations	Descriptions
$C_A = 140 \times A_{E\&C}$	Area cost (\$)
$C_{equipment} = 4 \times C_A$	Instrument cost (evaporator, condenser ...) (\$)
$C_{site} = 0.2 \times C_{eq}$	Site cost (\$)
$C_{tr} = 0.05 \times (C_A + C_{eq} + C_s)$	Transportation costs (\$)
$C_b = 0.15 \times C_{eq}$	Building costs (\$)
$C_b = 0.1 \times C_{eq}$	Engineers and salary costs (\$)
$C_c = 0.1 \times (C_A + C_{eq} + C_s)$	Contingency costs (\$)
$CC_{MED} = C_A + C_{equipment} + C_{site} + C_{tr} + C_b + C_b + C_c$	Capital costs (\$)

Table 6.
MED price [5].

which in these relations $A_{E\&C}$ is the total area of condenser and effects.

According to the model stated in the previous part, the maintenance cost can also be determined by calculating the cost of each component.

The operating cost of the reverse osmosis system is calculated as follows:

$$OCm = 0.2 \times CCm \tag{16}$$

$$OCinsurance = 0.005 \times TCI \tag{17}$$

$$OClaor = Qp \times 24 \times 365 \times fc \times 0.01 \tag{18}$$

$$OCmain = Qp \times 24 \times 365 \times fc \times 0.01 \tag{19}$$

$$OCch = Qp \times 24 \times 365 \times fc \times 0.0225 \tag{20}$$

$$OCO\&M, RO = OCinsurance + OClaor + OCmain + OCch \tag{21}$$

$$AOCRO = OCm + OCO\&M, RO \tag{22}$$

which in these equations OC_m is the replacement cost. $OC_{O\&M}$ is the total cost of operation, which includes OC_{laor}, OC_{main}, OC_{ch}, $OC_{insurance}$, which are the annual cost of the laboratory, the annual cost of repairs, the annual cost of chemicals and the insurance cost, respectively.

The operating cost of the MED system is calculated as follows [4];

$$C_{el} = c_{el} \times P \times f_c \times Q_p \times 365 \tag{23}$$

$$C_1 = 0.1 \times f_c \times Q_p \times 365 \tag{24}$$

$$C_{ch} = 0.04 \times f_c \times Q_p \times 365 \tag{25}$$

$$C_{in} = 0.005 \times C_A \tag{26}$$

$$AOC_{MED} = C_{th} + C_{el} + C_l + C_{ch} + C_{in} \tag{27}$$

which C_{el} is the cost of electricity, C_l is laboratory costs, C_{ch} is chemical costs, C_{in} is insurance costs, and finally AOC_{MED} is the annual operating cost. In relations (23) to (25), Q_p is equal to the flow rate of permeate water and P represents the power of the pumps. The operating cost of other components have also been calculated according to reference [3, 4]. Finally, the annual total of exploitation is calculated as follows:

$$AOC_{Total} = AOC_{Other} + AOC_{RO\&MED} \tag{28}$$

The normalized total cost is also determined from the eq. (29):

$$TAC = (TCI/CRF) + AOC_{Total} \tag{29}$$

Finally, the unit cost of fresh water production is calculated as follows.

$$UPC = \frac{TAC}{Qp \times 24 \times 365} \tag{30}$$

where in:
UPC: production cost of one m^3 of produced water ($/$m^3$).
TCC: Investment Cost ($).
AOC: annual operating cost ($/year).
OC: Operating Cost ($).
ΔP_{hpp}: High pressure pump pressure drop (MPa).
ΔP_{Tb}: Turbine pressure drop (Francis, Pelton) (MPa).
E_w energy consumption (KW).

Also, regarding the costs of other components of the cycle, we can refer to the suggested formulas of the reference which is stated below.

4.1 Gas turbine cycle cost

The gas turbine is made up of various components, the relationships related to the cost estimation of compressor (AC), combustion chamber (CC) and gas turbine (GT) can be expressed in the **Table 7**.

GT	$Z_{GT} = \left(\frac{a_{31} \dot{m}_g}{a_{32} - \eta_{GT}}\right) \ln\left(\frac{P_C}{P_D}\right)[1 + \exp(a_{33} T_C - a_{34})]$ $a_{31} = 479.34 \frac{\$.s}{kg}$ $a_{32} = 0.92$ $a_{33} = 0.036\ K^{-1}$ $a_{34} = 54.4$
CC	$Z_{CC} = \left(\frac{a_{21} \dot{m}_a}{a_{22} - \frac{P_C}{P_B}}\right)[1 + \exp(a_{23} T_C - a_{24})]$ $a_{21} = 46.08 \frac{\$.s}{kg}$ $a_{22} = 0.995$ $a_{23} = 0.018\ K^{-1}$ $a_{24} = 26.4$
AC	$Z_{AC} = \left(\frac{a_{11} \dot{m}_a}{a_{12} - \eta_{AC}}\right)\left(\frac{P_B}{P_A}\right) \ln\left(\frac{P_B}{P_A}\right)$ $a_{11} = 71.10 \frac{\$.s}{kg}$ $a_{12} = 0.9$

Table 7.
Gas turbine cycle costs.

4.2 Steam turbine cycle cost

The Steam turbine is made up of various components, the relationships related to the cost estimation of Steam turbine (ST), Pump (P) and Condenser (CON) can be expressed in the **Table 8**.

5. Input parameters for RO system modeling

In a practical process, several stages are used for the RO system, each stage includes several parallel Pressure Vessel (PV) that work with the same conditions. Each PV consists of several membrane elements connected in series. The feed water enters the first element and after purification, the condensed water (water coming out of the membrane) enters the second element and continues in the same way until the last element. The output of the products of all the elements are connected to each other and finally the water output of the final product is collected. The number of elements of the numerical series is between 2 and 8.

RO system modeling is done according to the entries in **Table 9**.

The assumptions used in this modeling are expressed as follows:

1. Constant consideration of temperature during the process.

2. Constant considering the permeability coefficients for water and different salts for each membrane [6–8].

5.1 Relevant equations for modeling an RO system.

At first, we should determine the average current intensity of the membrane (f) according to the type of water entering the membrane, then the total number of

ST	$Z_{ST} = a_1 w^{a_2} \phi_\eta . \phi_T$ $\phi_\eta = 1 + \left(\frac{1-\bar{\eta}_1}{1-\eta_1}\right)^{a_3}$ $\phi_T = 1 + a_4 . \exp\left(\frac{T_1 - \bar{T}_1}{a_5}\right)$ That: ϕ_η first law efficiency correction factor $\phi_T$correction factor for inlet steam temperature $a_1 = 3880.5 \frac{\$}{Kw^{0.7}}$ $a_2 = 0.7$ $a_3 = 3$ $a_4 = 5$ $a_5 = 10.42$ $\bar{\eta}_1 = 0.95$ $\bar{T}_1 = 866$ $\bar{\eta}_1 = 0.9$
P	$Z_{Pump} = a_1 (mV\Delta P)^{a_2} f_m . \phi_\eta$ $\phi_\eta = 1 + \left(\frac{1-\bar{\eta}_1}{1-\eta_1}\right)^{a_3}$ $f_m = \begin{cases} \text{Cast iron} = 1 \\ \text{Steel} = 1.41 \end{cases}$ $f_m = \begin{cases} \text{Cast iron} = 1 \\ \text{Steel} = 1.41 \end{cases}$ That: Fm f_m: correction factor of pipe material, in this case: f_m =1.41 ϕ_η: first law efficiency correction factor $a_1 = 549.13 \frac{\$}{Kw^{0.71}}$ $a_2 = 0.71$ $a_3 = 3$ $\bar{\eta}_1 = 0.8$
CON	$Z_{Con} = \frac{a_1 \dot{Q}_{Con}}{k.\Delta T_{in}} + a_2 \dot{m}_m + 70.5 \dot{Q}_{con} \times \left(-0.6936 \ln\left(\bar{T}_{cw} - T_b\right) + 2.1897\right)$ That: $a_1 = 280.74\ \$.m^{-2}$ $a_2 = 746\ \$.(Kg.s)^{-1}$ $k = 2200\ W.m^{-2}.K^{-1}$

Table 8.
Steam turbine cycle costs.

required elements of the N_E system is calculated by eq. (31) and the number of PVs is calculated from eq. (32) is determined. Using the tables in the DOW catalogs and the number of elements, we can determine the number of steps needed to achieve the desired recovery. By determining the number of stages using the stage ratio eq. (33), the number of pressure pipes in each stage is obtained eq. (34, 35) [4].

$$N_E = \frac{(Q_p)}{f \times S_m} \tag{31}$$

In this equation, Q_p is the permeate flow rate (m^3/h) and S_m is the membrane area (m^2).

$$N_V = \frac{N_E}{N_{E\ pv}} \tag{32}$$

Parameter	Unit
Inlet salt concentration	(mg/lit)
Feed water flow rate	(m^3/h)
Inlet water temperature	(°C)
Ambient pressure	(bar)
Number of PVs	(−)
Membrane type	(−)
Pure water permeability constant	(−)
Salt permeability constant	(kg/m^2.s.Pa)
Outer radius of the fiber bundle	(kg/m^2.s)
Inner radius of the fiber bundle	(m)
length of fiber bundle	(m)
Feed space	(m)
Pure water permeability constant	(m)
RO system stage number	(−)
The amount of total recovery of fresh water considered	(%)

Table 9.
Input parameters for RO system design.

N_v is the total number of pressure pipes and N_{Epv} is the number of series elements in each pressure pipe.

$$RR = \left[\frac{1}{1-R}\right]^{\frac{1}{n}} \tag{33}$$

RR is the step ratio, R is the system recovery and n is the number of steps.

$$N_{V(1)} = \frac{N_v}{1+RR^{-1}} \tag{34}$$

$$N_{V(2)} = \frac{N_v(1)}{1+RR} \tag{35}$$

$N_v(i)$ is the number of pressure tubes in the i-th stage.

According to the mass transfer relations, it can be seen that the flow rate of water and salt through the membrane will be in the form of eqs. (36) and (37) and also the average velocity in each element of the membrane in relation (38) will be determined [6–8]. The amount of salt concentration in produced water is determined from the eq. (39) [6]. Also, according to the concentration polarization phenomenon of the mass transfer process, the salt concentration near the wall is calculated based on the film theory in the form of eq. (40) [6]. According to the continuity equation, eqs. (41) and (42) can be used for water, and eq. (43) can be used for salt.

$$J_w = A \times TCF\left[\left(P_f - P_p - \frac{\Delta P_f}{2}\right) - (\pi_w - \pi_p)\right] \times 10^6 \tag{36}$$

$$J_s = B(C_w – C_p) \tag{37}$$

$$V_w = \frac{Jw + Js}{\rho_p} \tag{38}$$

$$C_p = \frac{Js}{Vw} \times 1000 \tag{39}$$

$$C_w = C_p \times \left[\left(P_f - P_p - \frac{\Delta P_f}{2}\right) - (\pi_w - \pi_p)\right] \times 10^6 \tag{40}$$

$$Q_p = V_w S_m \tag{41}$$

$$Q_B = Q_F - Q_P \tag{42}$$

$$C_B = \frac{Q_F C_F - Q_P C_P}{Q_B} \tag{43}$$

In the above equations, A and B and S_m is the permeability coefficient of water and salt in the membrane and its area, respectively. Permeation coefficients for different membranes are fixed and considered based on the Dow catalog. The relationships that can be used to reduce the number of adjectives are as follows [7–9];

$$k = 0.04 \times Re^{0.75} \times Sc^{0.33} \times \frac{D_s}{d} \tag{44}$$

$$\Delta P_f = \frac{0.0033 Q_a L_{PV} \mu}{Wd^3} \tag{45}$$

$$Q_a = \frac{Q_b + Q_f}{2} \tag{46}$$

$$\pi = \frac{0.2641 \times C(T + 273)}{1.0 \times 10^6 - C} \tag{47}$$

and in these relationships Re is Reynolds Number, D_s is the salt permeability (m^2/S), d is the distance of the feeds from each other, and Q_a is the average flow rate, which is calculated from the eq. (45). $L_{PV} = m.\ L_m$ where L_m is the length of the

Adjectives	**equation**
J_w, P_f, C_w, C_p	(36)
J_s	(37)
V_w	(38)
—	(39)
—	(40)
Q_p	(41)
Q_B	(42)
C_B	(43)

Table 10.
Adjectives in RO system modeling equations.

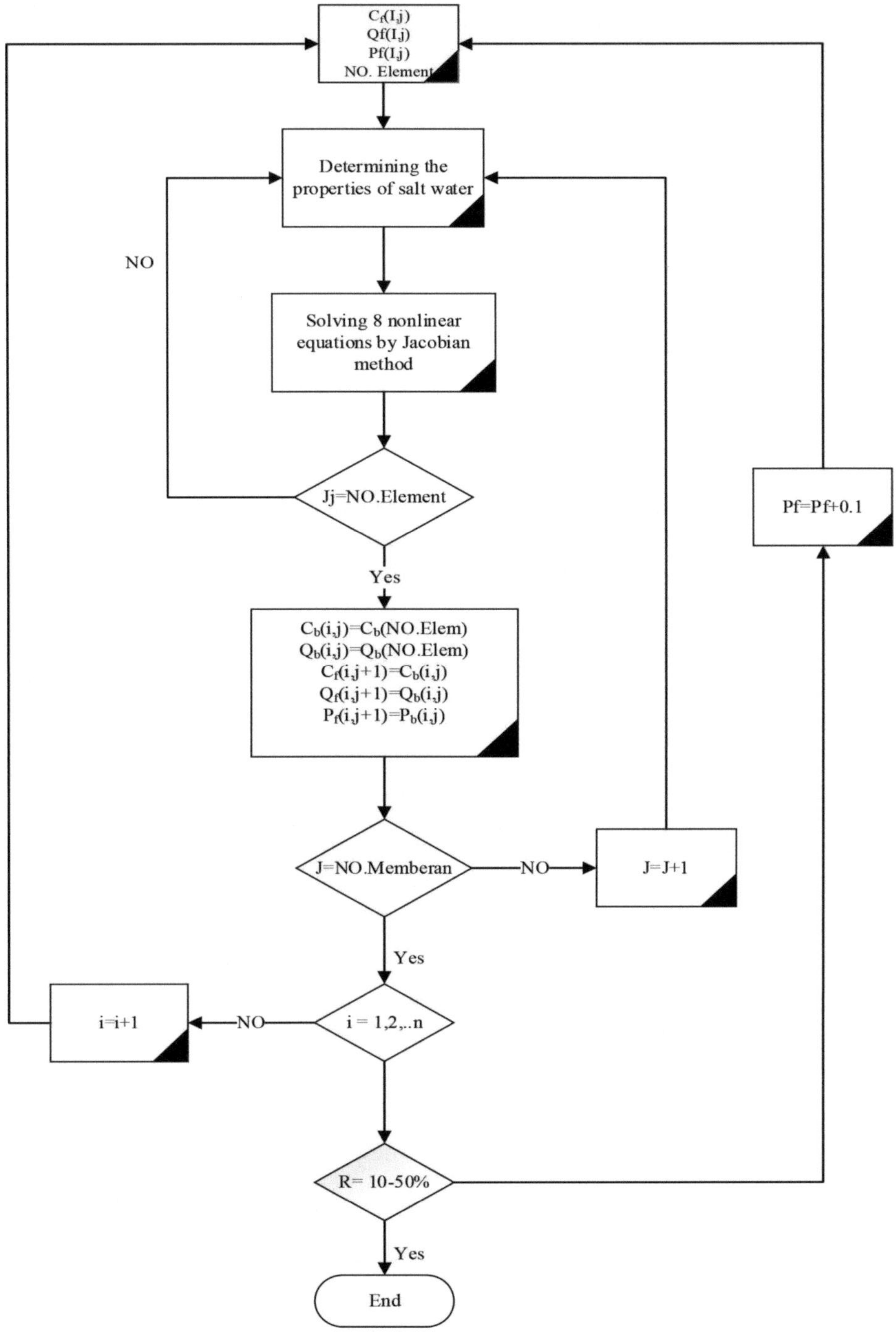

Figure 2.
Algorithm for solving equations in RO desalination system design.

membrane $\Delta P_f \leq 0.35\,MPa$. π is Osmotic pressure and C in this relationship is salt concentration.

In order to estimate the average pressure drop, the Hagen-Poisey equation is used. Sherrod's number is calculated according to the eq. (44) and (45); by which concentration polarization can be calculated.

For a spiral membrane element, each of the feed and product water flows can be considered as a flow between two parallel planes with length L, width W and distance d; and based on that, he calculated the pressure drop on the supply side. For the spiral element, the width of the membrane W can be calculated with the following relationship based on the area of the membrane and the number of sheets (N_l) $S_m = W \times l \times N_l$ [6]. According to the above equations from (36) to (43), the number of adjectives for each equation can be determined in the following **Table 10**. In some equations, the adjectives are repeated, we have tried to mention the adjectives that are expressed for the first time in the table in order to determine the number of variables completely.

As can be seen, there are 8 equations with 9 variables, which can solve 8 equations and 8 nonlinear adjectives by assuming the input pressure and correcting it.

5.2 RO system design flowchart

In some references, to solve these equations, the rate of salt rejection (for example, in Ref. [6]) or in some other references, the recovery of any RO system is assumed (reference [10]) and the equations are solved based on that. Nader et al. [8] presented another solution that solves equations with the same number of adjectives, which is based on the trial and error method. Here, due to the fact that an accurate and comprehensive modeling is used, the minimum assumptions of the method presented by Nader et al. [8], have been used, whose values can be seen in **Table 8**. But in the following, according to the determination of the inlet water pressure and the requirement to solve the equations by repetition method, the duration increases, and as a result, the equations are solved according to the solution of nonlinear equations by Newton's method. Finally, according to the repetition method, the amount of feed water inlet pressure has been determined by trial and error; According to the explanations given, the problem solving flowchart is presented in "**Figure 2**".

In the solution process, to increase the accuracy, each element is divided into smaller components, and the output from one component will be the input to another

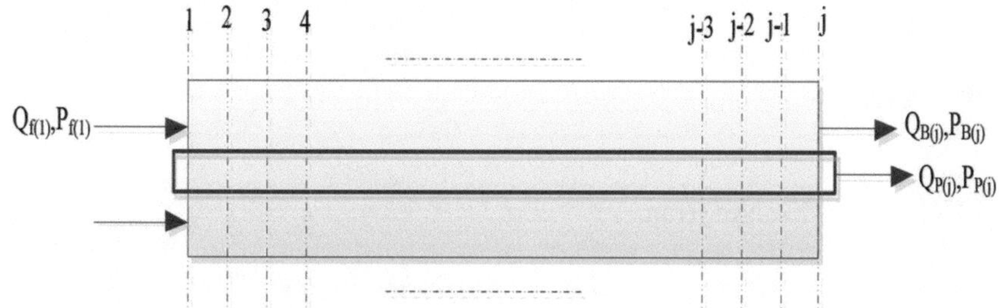

Figure 3.
Schematic of each element for RO system modeling [4].

component. "**Figure 3**" shows the method of dividing an element into smaller components to increase accuracy, which is chosen based on the method proposed by Nader et al. [8].

6. Conclusions

In this chapter, the technical and economic analysis of the reverse osmosis desalination system was discussed. Relationships and parameters that are important in discussing the cost of produced water in reverse osmosis desalination systems as well as different parts of the power plant were explained in detail. Also, relationships, parameters, and mathematical models used to simulate reverse osmosis water desalination were also explained.

Author details

Mehdi Sepahvand
Kashan University, Kashan, Iran

*Address all correspondence to: mehdi_s_1990@yahoo.com

References

[1] Haghparast HAM, Kabir K. Reducing the operating cost of reverse osmosis systems using hydro turbine energy recovery systems. In: The First International Conference of Power Plants; 15 April;. Iran; 2010

[2] Osmosis R. 2012. Available from: http://www.lenntech.com

[3] Bejan A, Tsatsaronis G, Moran M. Thermal Design and Optimization. Michigan: John Wiley and Sons; 1996

[4] Mokhtari H, Ahmadisedigh H, Ebrahimi I. Comparative 4E analysis for solar desalinated water production by utilizing organic fluid and water. Desalination. 2016;**377**:108-122. DOI: 10.1016/j.desal.2015.09.014

[5] Esfahani IJ, Ataei A, Shetty V, Oh T, Park JH, Yoo C. Modeling and genetic algorithm-based multi-objective optimization of the MED-TVC desalination system. Desalination. 2012; **292**:87-104. DOI: 10.1016/j. desal.2012.02.012

[6] Lu Y, Liao A, Hu Y. The design of reverse osmosis systems with multiple-feed and multiple-product. Desalination. 2012;**307**:42-50. DOI: 10.1016/j. desal.2012.08.025

[7] Du Y, Xie L, Liu J, Wang Y, Xu Y, Wang S. Multi-objective optimization of reverse osmosis networks by lexicographic optimization and augmented epsilon constraint method. Desalination. 2014;**333**(1):66-81. DOI: 10.1016/j.desal.2013.10.028

[8] Al-Bastaki NM, Abbas A. Predicting the performance of RO membranes. Desalination. 2000; **132**(1–3):181-187. DOI: 10.1016/ S0011-9164(00)00147-8

[9] El-Halwagi MM. Synthesis of reverse-osmosis networks for waste reduction. AIChE Journal. 1992;**38**(8):1185-1198. DOI: 10.1002/aic.690380806

[10] Maskan F, Wiley DE, Johnston LP, Clements DJ. Optimal design of reverse osmosis module networks. AIChE Journal. 2000;**46**(5):946-954. DOI: 10.1002/aic.690460509

Chapter 2

Household Water Treatment Practice

Dejen Tsegaye

Abstract

Improvements in water quality and a decrease in the prevalence of diarrheal disease in poor nations have been linked to household water treatment and safe storage practices. The objective of this study was to assess knowledge and practice of household water treatment and associated factors in rural kebeles of Dega Damot Woreda, North West Ethiopia, 2021. In Dega Damot Woreda, North West Ethiopia, in 2020, a community-based cross-sectional study was carried out. To choose 845 households in the study area, a multistage sampling procedure was used. Pretested questionnaires were used to collect the data, which was then entered into Epi-data for cleaning and analysis before being exported to SPSS, and multivariable logistic regression analysis was used to identify factors. Only 14% of participants in this research were actively treating their home's water, whereas 71.8% knew about the technique. The following variables were significantly associated with household water treatment practice: educational status, income earning >600ETB per month, number of children under five in the household, and methods of fetching water. In Dega Damot Woreda, there was severe lack of household water treatment practices. The Woreda health office needs to raise community awareness and knowledge of domestic water treatment techniques.

Keywords: household water treatment, knowledge and practice, factors, Dega Damot, Ethiopia

1. Introduction

A sufficient supply of clean water is one of the most fundamental human requirements and must be provided for as they are two of the most significant factors affecting public health [1]. Water that poses no major risk to health over the course of a lifetime is considered to be safe for drinking. The United Nations (UN) formally recognized the human right to access safe water without restriction in 2010. To sustain a population's excellent health, safe water is essential [2, 3]. It is common knowledge that having access to clean water and sanitary facilities helps to stop the spread of disease. Only having access to clean water does not greatly lessen diarrheal illnesses. Even if the source is clean, feces can contaminate water during collection, transportation, storage, and home drawing [4–6].

Prior to usage, drinking water is subjected to household water treatment (HWT), which enhances its microbiologic purity. Due to the possibility of recontamination during the process of transport, storage, and consumption, it is thought to be superior to treatment at other levels (such as the source). It has been demonstrated to be among the most efficient and economical methods of preventing waterborne illnesses. Therefore, vulnerable groups take charge of their water security by treating and storing household water safely [7–9].

HWT can enhance the quality of drinking water at the point of use and lower the risk of diarrhea in the millions of people who rely on improved and unimproved water sources. HWT includes boiling, chlorination, filtration, and solar disinfection. When populations at risk of waterborne disease adopt efficient HWT procedures appropriately and consistently, the risk of diarrheal disease can be reduced by as much as 61% [10–12].

The majority of the world's 1.8 billion users of fecally contaminated water sources are in low- and middle-income nations. The largest health concern associated with water consumption is microorganisms found in water that has been feces-contaminated [13].

Nowadays, simple, low-cost, and acceptable household water treatment technologies are available. However, in many communities, there is limited knowledge and poor practice for water treatment [14]. Limited knowledge, misinformation, and lack of experience in best practices of alternative water treatment technologies are among the leading challenges [15]. People are not always aware of the risk related to transportation practices, storage, and handling of drinking water.

Nearly 90% of Ethiopia's rural residents do not use alternate water treatment techniques, putting them at significant risk for disease unless quick action is taken, such as alternative HWT techniques with safe water storage [16]. Furthermore, there are few studies on HWT knowledge, behaviors, and related factors in Ethiopia. Therefore, the purpose of this study is to evaluate household water treatment knowledge and practice in rural kebeles in Dega Damot Woreda, North West Ethiopia. The town/urban areas of eastern Ethiopia were where the majority of studies were conducted. The purpose of the current study is to evaluate home water treatment knowledge and practice in the study area.

2. Methods

2.1 Study area and period

The West Gojjam Zone's Dega Damot Woreda is where this study was carried out. The distance between Bahir Dar City, the seat of the Amhara Regional State, and Addis Ababa, the capital city of Ethiopia, is 275 kilometers. The district has a 41% highland climate, a 37% temperate climate, and a 22% lowland climate. In 2019, it will have an expected 184,369 residents (91,263 men and 93,106 women), who will be split among 42,877 houses. More than 99% of followers are orthodox. There are two urban and thirty-two rural Kebeles, seven health centers, one general hospital, two private clinics, and one private pharmacy [17]. There are 779 functional and 20 nonfunctional hand-dug wells, 68 functional and four nonfunctional protected springs, and two functional and one nonfunctional borehole. The rural population who use protected water sources is 138,740 (82.4%) [17]. The study was conducted from March 20/2021–April 20/2021.

Study design: Community-based cross-sectional study was employed.

Source population: All households in rural kebeles of Dega Damot Woreda.

Study populations: All households in the selected rural kebeles of Dega Damot Woreda.

Study subjects: Mothers who live in selected rural kebeles of Dega Damot Woreda.

Inclusion criteria: Mothers in the household were included in the selected kebeles.

Exclusive criteria: Mothers in the household who resided for less than 6 months were excluded from the study.

Dependent variables: Knowledge and practice of HWT.

Independent variables.

Sociodemographic characteristics: Sex, age, educational status, family size, marital status, occupation, religion, household income, and ethnicity were dependent variables that are found under sociodemographic character.

Knowledge about HWT: Knowledge of HWT methods, knowledge of purpose HWT, knowledge of water born disease, knowledge of negative outcome of drinking dirty water, and knowledge of causes and prevention of diarrhea.

Water source and handling status: Source of drinking water, type of container to fetch water, distance to fetch water, type of container to store water, and way of fetching water from container.

Operational definition.

Knowledge: Respondents are able to identify methods of HWT, recognize the importance of treating drinking water, and identify diseases that can result from drinking unclean water. Variables in the questionnaire were given a total score ranging from 0 to n where n is the number of knowledge questions. Using frequency distribution, a score of <50% of the total knowledge questions was considered as poor knowledge, whereas a score of ≥50% of the total knowledge questionswas labeled as good knowledge [15].

Household water treatment practice: Households who used at least one alternative method of HWT within the last 24 hrs were considered as good practices, which will be scored as one, while poor practices were considered as households who were not using any alternative method of HWT and scored as 0 [15].

Sample size determination and procedure.

Single population proportion formula was used to determine sample size with assumptions of 5% margin of error (d) 95% CI (Z = 1.96), design effect (d) of 2 and 10% nonresponse rate and taking prevalence of practice 44.8% from the study done in Burie, Northwest Ethiopia [18]. Thus, the final sample size was 845. A multistage sampling technique was used. Twenty percent of kebeles in Dega Damot Woreda were selected by simple random sampling method. The samples were distributed proportionally by the number of households for each selected kebeles. Study participants were selected by systematic random sampling from HHs in the selected kebeles. The sampling interval (k) was determined by study population (5218 HHS in the selected kebeles) divided by sample size (845) =6. Then, the data were collected at every six HH intervals. Lottery method was used to select the first study subject. Respondents were mothers of the households. In case, if there were more than one mother in the household, one of them was selected by lottery method.

2.2 Data collection tools and procedures

Socio-demographic characteristics were collected through face-to-face interviews and observation with mothers. The questionnaire and observation checklist were developed in English and were translated into local language (Amharic) and were translated back to English to keep the consistency prior to the actual data collection. Data were collected by ten students who completed grade 12 and were supervised by two public health officers.

2.3 Data quality control

The questionnaire was pretested on 5% of the sample size to check understandability and reliability of the questionnaires. One-day training was given to data collectors and supervisors on the study instrument, data collection procedure, and the ethical principles of confidentiality. The collected data were reviewed and checked for completeness and relevance by the supervisors and principal investigator each day.

2.4 Data processing and analysis

The questionnaire was manually reviewed for accuracy. It was afterward coded, inputted into Epi-Data version 4.2, and exported to SPSS version 25 for additional analysis. The population was explained using descriptive statistics in relation to the pertinent variables. Chi-square testing was conducted. The multivariable logistic regression was fitted to the variables with fewer than 0.25 p-values from the bivariate analysis using the binary logistic regression technique. In the multivariable logistic regression, odd ratios with 95% confidence intervals (CI) were generated, and statistical significance was assessed at p-values 0.05. Hosmer and Lemeshow tests were used to assess the fitness of the models. Text, tables, and graphs were utilized to present the data.

2.5 Ethical consideration

Ethical clearance was obtained from the ethical committee of BDU College of Medicine and Health Sciences and a letter of cooperation was delivered to the Dega Damot Woreda administration bureau in order to get letter of permission for kebeles. Anyone who has no willingness to participate in the study was not forced to participate. Informed (verbal) consent was obtained from each study participant. The study participants were also provided with information about the objectives and expected outcomes of the study. Information obtained from individual participants was kept secure and confidential.

3. Results

3.1 Sociodemographic characteristics of participants

This study included 845 mothers in all, with a 100% response rate. The respondents' mean (+SD) age was 40.46 (+12.16) years, and 64.9% of them were illiterate. Respondents had a mean (+SD) family size of 4.88 (+1.2). Nearly all of the interviewees were farmers and Christians, and most (87.2%) were married. More than half of the households made monthly incomes of over 600 ETB, (**Table 1**), (**Figure 1**).

3.2 Practice of respondents on HWT

Only 14.1% of the 845 participants were using HWT. For storing drinking water, nearly half of the respondents (51.7%) had two containers; the remaining respondents had three (27.7%), one (15.4%), and four or more (5.2%), respectively. Nearly all (98.8%) of the responders possessed a container large enough to hold more than 25 liters. Total of 43.6% of respondents reported fetching drinking water three times daily, while the rest of respondents did so only twice, three times, or more, once, or only once. The majority (96.9) of the household's drinking water storage containers were plastic containers (rotto). Others utilized iron containers (0.4%) and clay pots (2.72%). Similarly, they used jerican (96.8%) and clay pots for the remaining portion of water retrieval. Nearly all families (98.7%) had clean household water containers, and of those, little under half (53.8) were cleaned once a week. The others were cleaned every day (11.5%) and within three days (34.7%) (**Table 2**), (**Figures 2** and **3**).

3.3 Knowledge of participants on HWT

About 28.2% of households have adequate knowledge of HWT, which is close to one-third. Only 24.4% of the respondents acknowledged knowing at least one HWT method, with the remainder not having done so. And of those, 84.4% had mentioned boiling, while the rest were familiar with chlorine. Only 34.8% of homes answered "yes" to the question "is it advisable to treat water for promoting child health" since the majority (84%) of households had little knowledge of diseases that are transmitted by water (**Table 3**), (**Figures 4–6**).

Variables	Response	Frequency (n = 845)	Percentage (100)
Educational status	Unable to read and write	548	64.9
	Read and write	297	35.1
Family size	< 5	609	72.1
	>5	236	27.9
Marital status	Married	737	87.2
	Single	33	3.9
	Divorced	25	3
	Widowed	50	5.9
Income	< 600 ETB	285	33.7
	≥ 600 ETB	560	66.3
Occupation	Farmers	843	99.8
	Others	2	0.2
No. of under-five children	No under-five children	450	53.2
	One,	342	40.5
	Two, and above	53	6.3

Table 1.
Sociodemographic characteristics of participants in Dega Damot Woreda, North West Ethiopia, 2021 (n = 845).

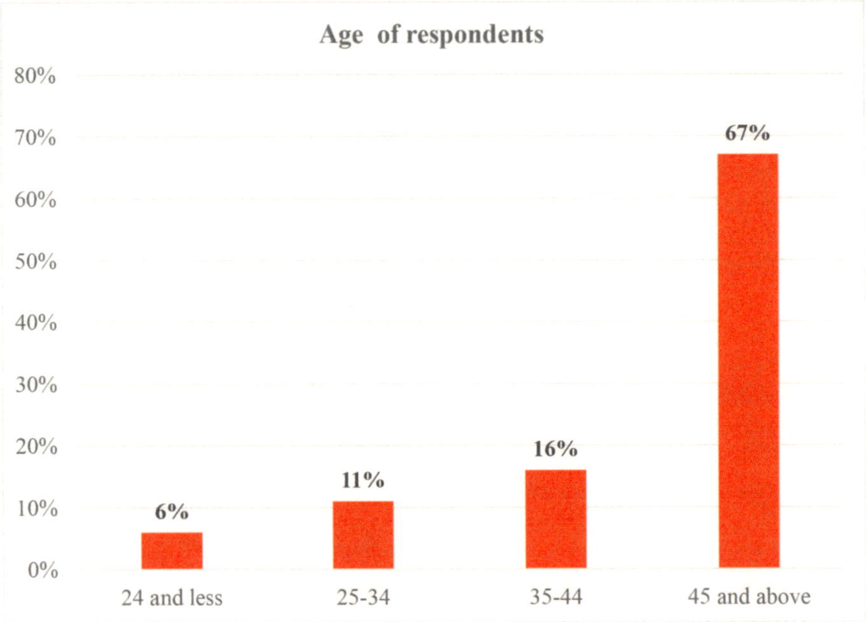

Figure 1.
Age of respondents in Dega Damot Woreda selective kebeles, North West Ethiopia, 2020 (n = 845).

Variables	Category	Frequency	Percent (100)
Overall HWT practice	Yes	119	14.1
	No	726	85.9
Individuals who fetch water	Mother	579	68.5
	Daughter	257	30.4
	Son	9	1.1
Distance to fetch water	<30 minute	643	76.1
	30–60 minute	192	22.7
	>60 minute	10	1.2
Number of days the water stored in the HH	One,	749	88.6
	Two,	74	8.8
	Three, and above	22	2.6
Days of the week water source has no service	Yes	39	4.6
	No	806	95.4
Way of fetching water from the containers	Pouring	467	55.3
	Dipping	378	44.7

Table 2.
Practice of respondents on HWT in Dega Damot Woreda selective kebeles, Amhara, Ethiopia, 2021 (N = 845).

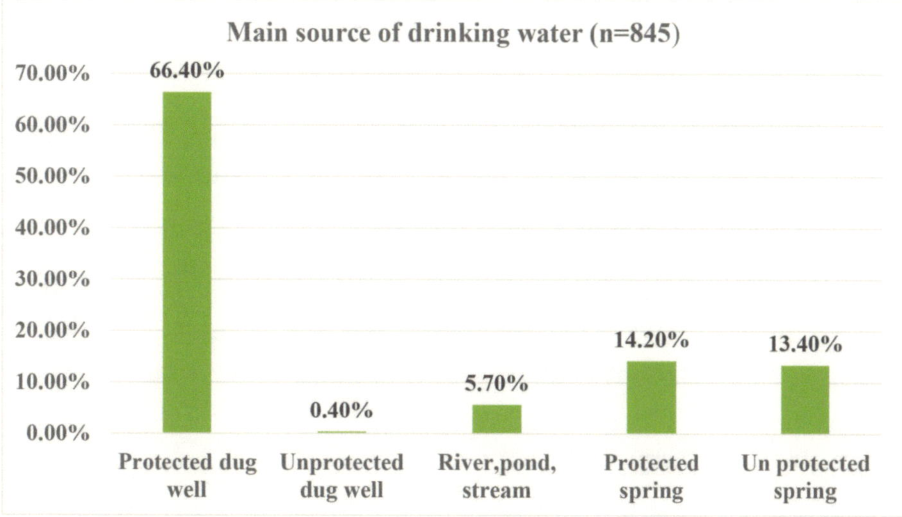

Figure 2.
Main source of drinking water for respondents in Dega Damot Woreda selective kebeles, North West Ethiopia, 2021 (n = 845).

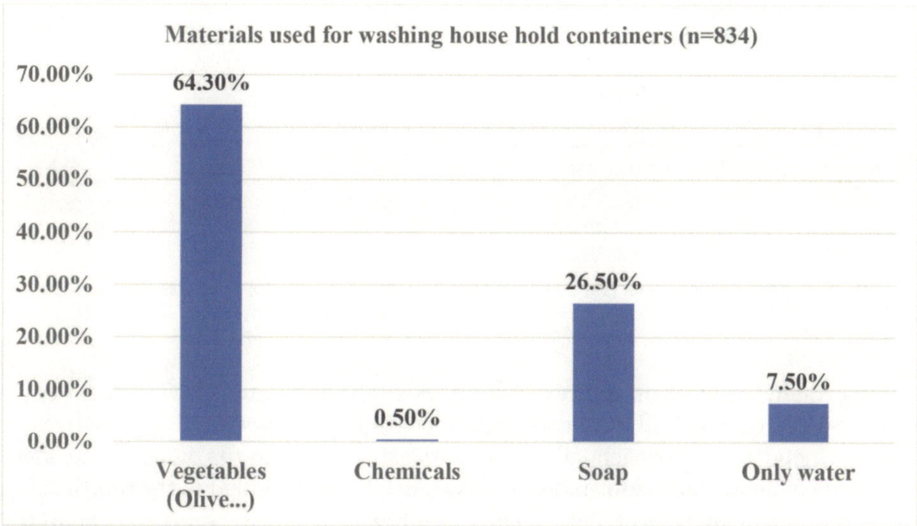

Figure 3.
Materials used for washing household containers for respondents in Dega Damot Woreda selective kebeles, Amhara, Ethiopia, 2021 (n = 845).

3.4 Bivariate and multivariate analysis of factors associated with practices on HWT

Age, educational level, family size, income, the number of children under the age of five, the method used to obtain drinking water, the type of container used to store

Variables	Response	Frequency	Percent (100%)
Overall Knowledge of participants on HWT	Good	238	28.2
	Poor	607	71.8
Knowledge about any disease caused by dirty water	Yes	194	23
	No	651	77
Childhood diarrheal disease prevented by safe water	Yes	307	36.3
	No	538	63.7
Knowledge about contamination of water at HH level	Yes	423	50.1
	No	422	49.9
Drinking water contaminated by unclean drinking utensils	Yes	467	55.3
	No	378	44.7
Difference between protected and unprotected water source	Yes	166	16.9
	No	679	84.1
Protected water sources may not be completely free from pathogenic organisms	Yes	155	18.3
	No	690	81.7
Cleanliness of drinking water by necked eye only	Yes	395	46.7
	No	450	53.3
Narrow necked water container is better than wide necked to prevent water contamination	Yes	591	69.9
	No	254	31.1
Children are more susceptible to diarrheal disease	Yes	375	44.4
	No	470	53.6
Treated water intake reduces family medical expense	Yes	347	41.1
	No	498	58.9

Table 3.
Knowledge level of the respondents on HWT in Dega Damot selective Woreda, Amhara, Ethiopia, 2021 (N = 845).

drinking water, the location where drinking water handling utensils were handled, and knowledge of HWT all had an association with HWT practice. Using the backward likelihood ratio approach, all factors with associations to the outcome variables in bivariate logistic regression analyses (p-value 0.25) were added to the multivariate logistic regression analysis models. Then, in multivariate logistic regression analysis, parameters such as educational status, income, the number of children under the age of five, the methods used to obtain drinking water, and HWT knowledge were found to be substantially associated with the practice of HWT.

The odds of practicing the HWT are more than seven times higher in homes with literacy than in households without literacy [AOR: 7.27, 95% CI: (4.36–12.11)]. When compared to households earning less than 600 ETB per month, households earning more than 600 ETB per month are almost three times more likely to practice HWT [AOR: 2.71 95% CI: (1.45–5.05)]. When compared to households with two or more under-five children, those without under-five children had an 83% lower likelihood of practicing HWT (AOR: 0.17, 95% CI: (.07–.41)). Similar to this, households that used pouring to obtain drinking water from the container are 0.42 times less likely to engage in HWT than households that utilized dipping [AOR: 0.42 95% CI: (.26–.67)].

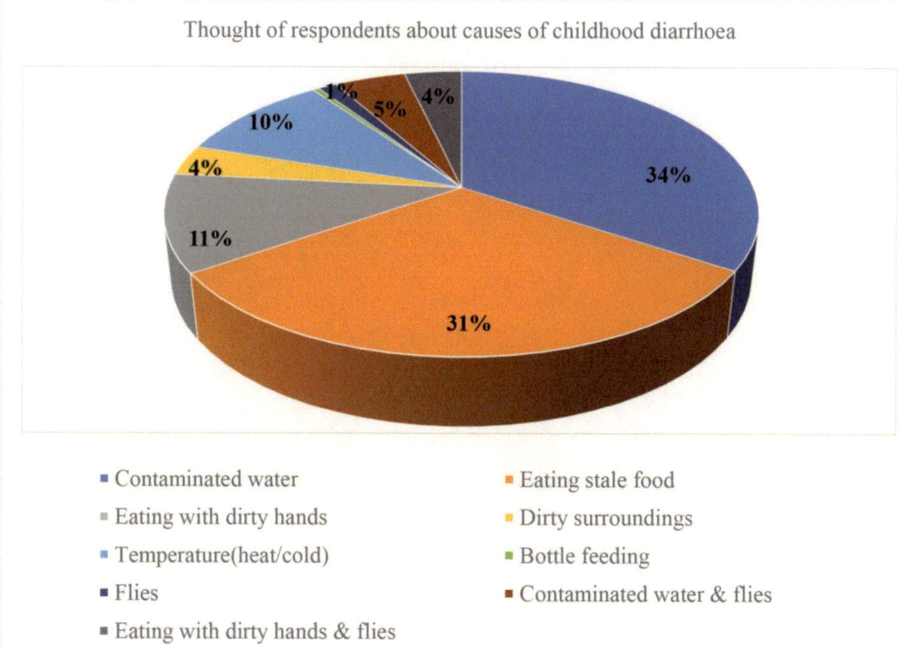

Figure 4.
Thoughts of respondents about causes of childhood diarrhea in Degad Dmot Woreda selective kebeles, Amhara, Ethiopia, 2021 (n = 845).

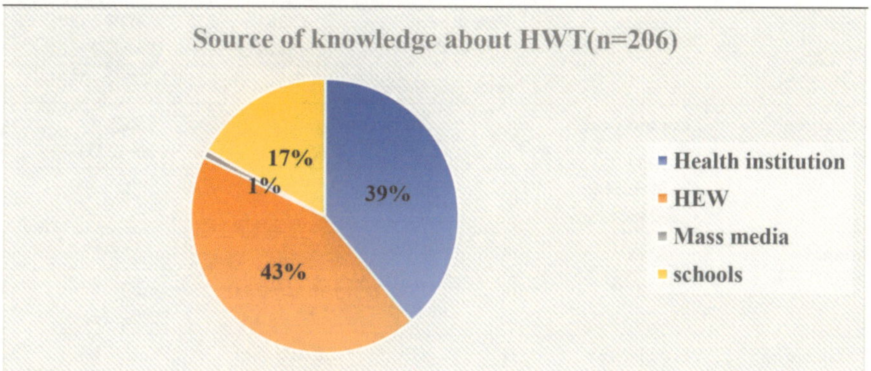

Figure 5.
Source of knowledge for respondents in Dega Damot Woreda selective kebeles, Amhara, Ethiopia, 2020 (n = 845).

Additionally, compared to their counterparts, those who had solid knowledge of HWT were approximately three times more likely to practice it [AOR: 3.03, 95% CI (1.84–5.01)] (**Table 4**).

3.5 Bivariate and multivariate analysis of factors associated with knowledge of HWT

Binary logistic regression was used to find the variables connected to HWT knowledge. Age of respondents, educational level, marital status, income, source of water to fetch, quantity of containers to fetch, methods to fetch drinking water, type of

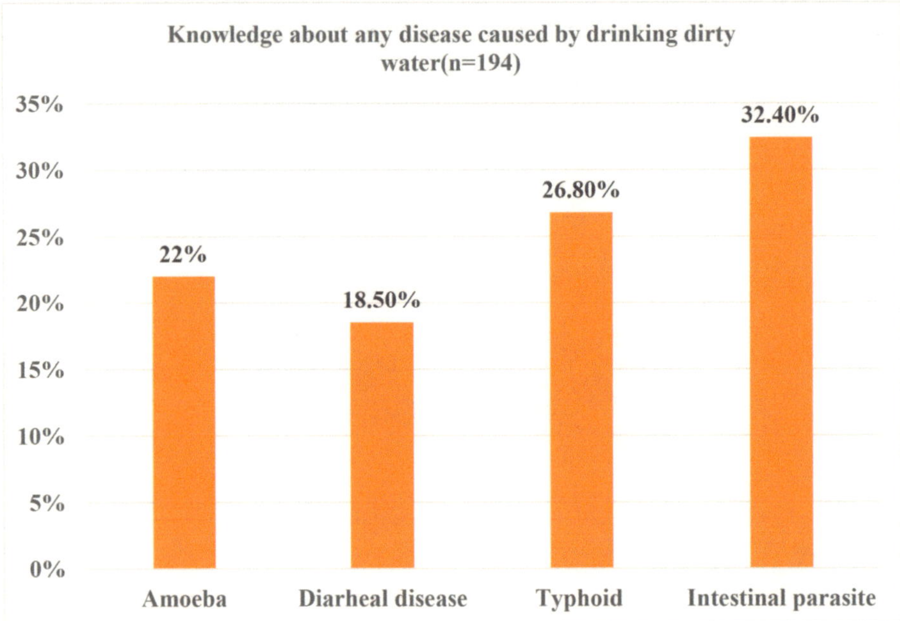

Figure 6.
Knowledge about any disease caused by drinking dirty water of respondents in Dega Damot Woreda, North West Ethiopia, 2021.

Variables	Category	Practice on HWT		COR (95%CI)	AOR (95%CI)	P-value
		Yes	No			
Educational status	Read and write	90	207	7.78 (4.96–12.18)	7.27 (4.36–12.11)	000
	Unable to read and write	29	519	1	1	
Income	<600	14	271	1	1	
	≥600	105	455	4.46 (2.5–7.95)	2.71 (1.45–5.05)	.002
Number of <5 children	No under-five children	21	429	.18 (.08–.41)	.17 (.07–.41)	000
	1	87	255	1.3 (.64–2.64)	.79 (.36–1.75)	
	≥2	11	42	1	1	
Ways to fetch water	Pouring	44	423	.42 (.28–.62)	.42 (.26–.67)	000
	Dipping	75	303	1	1	
Knowledge on HWT	Good knowledge	47	191	1.83 (1.22–2.74)	3.03 (1.84–5.01)	000
	Poor knowledge	72	535	1	1	

Table 4.
Bivariate and multivariate analysis of factors associated with practice on HWT among respondents in Dega Damot selective kebeles, Amhara, Ethiopia, 2021 (n = 845).

container to store drinking water, and location of handling utensils for drinking water all had associations with knowledge of HWT practice in bivariate logistic regression analysis. Using the backward likelihood ratio approach, all factors from the bivariate logistic regression analyses that have a relationship with the outcome variables were incorporated into the multivariate logistic regression analysis models. The factors that were significantly associated with knowledge of HWT in the multivariable logistic regression analysis were educational level, marital status, source of drinking water, number of containers for drinking water (those who had two and three or more), and locations to handle drinking water utensils.

The odds of having knowledge of the HWT are 1.78 times greater in households with literacy than in households without literacy [AOR: 1.784, 95% CI: (1.237–2.572)]. Being single increases the likelihood of knowing about HWT compared to households with widows [AOR: 4.68, 95% CI: (1.68–13.05). Similar to this, families with protected drinking water sources have nearly three times the likelihood of knowing about HWT than those with unprotected sources [AOR: 2.73, 95% CI: (1.88–3.96)]. In this regard, the odds of having knowledge of HWT are nearly two times higher in households with two water storage containers than in households with only one container

Variables	Category	Knowledge on HWT		COR (95% CI)	AOR (95%CI)	P-value
		Good	Poor			
Educational status	Read and write	174	374	1.69 (1.42–3.82)	1.78 (1.23–2.57)	000
	Unable to read and write	64	233	1	1	
Marital status	Married	200	535	1.06 (.55–2.03)	1.24 (.62–2.47)	
	Single	15	18	2.37 (.93–6.02)	4.68 (1.68–13.05)	000
	Divorced	10	15	1.89 (.68–5.26)	2.73 (.94–7.92)	
	Widowed	13	37	1	1	
Source of water to fetch	Unprotected	72	92	2.42 (1.7–3.46)	2.73 (1.88–3.96)	000
	Protected	166	515	1	1	
No of containers to store water	One	192	562	1	1	
	Two	27	37	2.13 (1.26–3.6)	2.22 (1.29–3.84)	0.04
	Three and above	19	8	6.95 (2.99–16.13)	7.59 (3.15–18.27)	000
Place of handling utensils	On shelf (over the floor)	141	271	1.8 (1.33–2.44)	1.86 (1.34–2.56)	000
	Anywhere on the floor	97	336	1	1	

Table 5.
Bivariate and multivariate analysis of factors associated with knowledge of HWT among respondents in Dega Damot selective kebeles, Amhara, Ethiopia, 2021 (n = 845).

[AOR: 2.22, 95% CI: (1.29–3.84)] and nearly eight times higher in households with three or more water storage containers than in households with only one container [AOR: 7.59, 95% CI: (1.29–3.84)]. Additionally, the likelihood that a family handles drinking utensils on a shelf as opposed to handling them randomly on the floor is nearly twice as high [AOR: 1.86, 95% CI: (1.34–2.56) (**Table 5**).

4. Discussion

Water is the most significant factor affecting public health, and having access to enough clean water is crucial for lowering disease transmission. Access to clean water does not dramatically reduce disease rates even if the source is safe since it can become faecally polluted during collection, transit, storage, and drawing in the home [4–6]. Above all, it is crucial to be knowledgeable about household water treatment and to put that information into practice by using highly advised techniques.

According to this study, HWT practice was found to be 14.1% (CI 11.8–16.3). This self-reported study's prevalence of HWT practice (14.1%) was much lower than studies carried out in India (53%), Zambia (50%), Nigeria (54%), and Kenya (69%) [9, 19–21], respectively. The disparity may result from different coverage of clean water as well as different household-level water treatment options across the nation depending on people's knowledge of the availability and quality of water. Additionally, Ethiopian communities, particularly in rural regions, do not use this water purification procedure [16].

This study's results were lower than those of a study done in North West Ethiopia (23.1%), as well. The discrepancy is likely the result of a different study environment where the community in the prior study received information from many sources and, as a result, had greater awareness of the problem than the study site in the present [15]. The current finding, however, was slightly higher than a study carried out in a rural area of Haryana, India (10%) [22]. The difference could be the result of a time difference between now and seven years ago when the prior was completed. Additionally, the sample size used in the earlier study was less than half of the sample size employed in this investigation.

When examining the extent of HWT knowledge, it was discovered to be 28.2% CI (25.3–31.5). This result was consistent with a research carried out in Nigeria (26.1%) [9]. However, this was considerably less than research conducted in India (69%) [19]. The original study was carried out in a nation that is more developed than the current study area, where it would have been possible to provide information regarding HWT that was more easily accessible. Additionally, this was less than the research conducted in North West Ethiopia (49.3%) [15]. This discrepancy may be the result of the communities' varying socioeconomic conditions, which may have an impact on how they use source water for drinking. However, it exceeds a research conducted in Patan (16.7%) [23]. This is most likely a result of the use of a tiny sample size, which is almost one-fourth of the sample size used in the current study.

There were variables in this study that showed a strong correlation with HWT practice. The first one was the level of education in each household. Reading and writing-capable households performed HWT better than their counterparts. Two studies conducted in Ethiopia's Bure Zuria and Dabat districts backed up the conclusion [15, 18]. This is because literate people are better able to learn about HWT practice and comprehend procedures than their illiterate counterparts.

The second factor that was substantially linked to practicing the HWT was having a household income of more than 600 ETB per month, which was 2.71 times higher than that of their counterpart. This was corroborated by a study carried out in North West Ethiopia, which explained that the more money a household makes, the more they can afford to purchase the supplies required for therapy [15].

Thirdly, HWT was less common in homes with less than five kids compared to those with just one. Since this study indicated that most households (52.4%) are aware that untreated water causes juvenile diarrheal disease, it is probably because mothers who live in households with children practice HWT more to protect their children from water-borne illness. The fourth substantially linked variable was the likelihood of HWT practiced by households; these households were less likely to obtain their drinking water by pouring from the container. This might be because participants believed that pouring was a secure way to handle water.

Good knowledge of HWT practice is the final and fifth factor that is significantly related to HWT practice. Research conducted in Patan and North West Ethiopia supported this [15, 23]. The more information families have about HWT, the more likely they are to use it.

Knowledge of HWT was another dependent variable in this study. The first factor that was strongly linked to this variable was educational attainment. Reading and writing-capable households were more likely to be aware of HWT. A study conducted in Patan, Biye Kaduna state, Nigeria, and Dabat North West Ethiopia provided evidence in favor of this [9, 15, 23, 24]. It goes without saying that being able to read and write is crucial if one wants to increase their knowledge through various methods.

The second variable that was significantly linked to HWT knowledge was marital status. Single-person households knew more about HWT than widowed households did. This is supported by a study done in Patan [23]. Due to the lack of children or elderly people to carry out the practice, singletons are likely to have a lighter workload. Additionally, singles had higher levels of education than divorced people (88% vs. 12%).

Thirdly, factors related to understanding of HWT included sources of water that were protected. Families with improved/protected drinking water sources have higher levels of knowledge than their counterparts. This was corroborated by a study conducted in Northwestern Ethiopia [15]. This suggests that households take more precautions to avoid using unprotected drinking water the more they are aware of HWT. The knowledge of all the negative effects of unprotected water on health also made people aware of the need to use protected water sources.

The fourth and final variable was the number of water storage containers, and it was substantially correlated with understanding of HWT. Homes with two water storage containers for drinking water were more likely to be aware of HWT than homes with just one container. Similar to this, homes with three water storage containers were more likely to be familiar with HWT than those with just one container, and even they were more familiar with it than households with only two containers. Households on the HWT may already be aware of this, and its benefit may have forced them to have more water storage tanks. The number of water bottles a household has actually indicated how well-versed they are in using them individually for various functions. The others may be used for different purposes, while one may be used for dipping water that is obtained from containers by fixing it within.

Last but not least, the location where drinking utensils were handled was a factor that significantly correlated with HWT knowledge. Families who handled their utensils on a shelf or anywhere other than the floor were more likely to be familiar with HWT than those whose utensils were handled on the floor. This suggests that handling their utensils while on the shelf, on the floor, or in a safe place may protect homes from many water-borne diseases. And they are acting in this way because they are aware of proper utensil handling. This is the fact that all water drinking supplies should be stored safely and away from any unclean items, such as on a shelf or somewhere else other than the ground.

5. Conclusion and recommendations

5.1 Conclusion

According to this study, there is a lack of HWT practice and knowledge in the Dega Damot Woreda. Factors substantially linked with HWT practice included educational status, income earning >600ETB, the number of children in the home under the age of five, the means of fetching water, and understanding of HWT. In contrast, characteristics such as educational level, marital status, drinking water source, quantity of water storage containers, and location of utensil handling exhibited a significant association with understanding of HWT.

5.2 Recommendations

The author offers the recommendations below in light of the findings of this study:

Woreda government office: The Woreda office, working with the Woreda health office, shall provide protected water for drinking in order to raise knowledge of the regional state.

Dega Damot Woreda water office: It is better to inform the community about HWT procedures and show them by kebeles/sub kebeles when the Woreda water office collaborates with the Woreda health office. Additionally, they must demonstrate how to obtain and use the chemicals used in water treatment.

Nongovernmental organizations: Nongovernmental organizations that are involved in the water supply are better to perform wonderful activities to enhance community knowledge and their practice. It is also preferable to adopt the supporting resources required for HWT practice as soon as the community approached to do so.

Author details

Dejen Tsegaye
Department of Nursing, College of Health Nursing, Debre Markos University, Ethiopia

*Address all correspondence to: dejenetsegaye8@gmail.com

References

[1] SCA SCoA. The microbiology of drinking water. Part 1-water quality and public health methods for the examination of waters and associated materials. Environment Agency. 2002

[2] Vanessa G. Household Water Treatment and Safe Storage Options for Northern Region Ghana. 2008

[3] WHO. A Guide to Equitable Water Safety Planning. World Health Organization; 2019

[4] Amira ME-AA, Sulieman ME, el-Khalifa EA. Microbiological assessment of drinking water quality in wadmedani & Khartoum states, Istanbul, Turkey. In: Sixteenth International Water Technology Conference. 2012. pp. 1-13

[5] Divekulu SD, Woyessa. Assessment of bacteriological quality and traditional treatment methods of water-borne diseases among well water users in Jimma town, southwest Ethiopia. Agricultural and Biological Science. 2013;**8**(6):477-486

[6] Teferi A. Assessment of Knowledge and Hygeinic Practices towards Bacteriological Quality of Drinking Water At Dobe Toga Kebele. Shebedino Woreda; 2007

[7] Kendralyn GJ. A Survey of Point of Use Household Water Treatment Options for Rural South India Public Health. 2012

[8] Simonne R, Mäusezahl D, Hans-Joachim M, Weingartner R. Quality of drinking-water at source and point-of-consumption–drinking cup as a high potential recontamination risk: A field study in Bolivia. Health Population Nutrition. 2010;**28**(1):34-41

[9] Miner CD, Zoakah A, Afolaranmi T, Envuladu EA. Household drinking water; knowledge and practice of purification in a community of Lamingo, Plateau state, Nigeria. Environmental Research and Management. 2015;**6**(3):230-236

[10] Bea R. Global assessment of exposure to fecal contamination through drinking water based on a systematic review. Tropical Medicine and International Health. 2014;**19**(8):917-927

[11] Haji A, Monica M. Knowledge, attitude and practice on solar water disinfectant at house hold water treatment in Maalim salat location, Wajir county. International Journal of Public Health. 2017;**1**(2):12-26

[12] Organization WH. Results of Round II of the WHO International Scheme to Evaluate Household Water Treatment Technologies. Geneva. Licence: CC BY-NC-SA 30 IGO; 2019

[13] Kristen H. Strategies for the Promotion of Household Water Treatment in Ica. Peru: Public Health-Epidemiology; 2015

[14] Abraham GB, Jonathan M, Daniele S, Esayas A, Sahilu G. Appropriate household water treatment methods in Ethiopia: household use and associated factors based on 2005, 2011, and 2016 EDHS data. Environmental Health and Preventive Medicine. 2018;**23**(46)

[15] Destaw B, Kebede Y, Andargie G, Tadesse T. Knowledge, attitude, and practice of mothers/caregivers on household water treatment methods in Northwest Ethiopia: A community-based cross-sectional study. American Society of Tropical Medicine and Hygiene. 2017;**2017**:97

[16] Calverton M. Ethiopian Demographic and Health Survey 2011. Ethiopia Central Statistical Agency and ICF International: Addis Ababa; 2012

[17] DDWHO. Annual report organized by Dega Damot worda health office. unpublished document. 2019

[18] Belay H, Dagnew Z, Abebe N. Small scale water treatment practice and associated factors at Burie Zuria Woreda rural households, Northwest Ethiopia, 2015: Cross sectional study. BMC Public Health. 2016

[19] Satapathy M, Subrat KPU, Swain DP, Mishra RP. Assessment of household water treatment and storage practices in India. Community Medicine and Public Health. 2018;**5**(3):1060-1063

[20] Ghislaine RP, Thomas C. Consistency of use and effectiveness of household water treatment practices among urban and rural populations claiming to treat their drinking water at home a case study in Zambia. Tropical Medicine and Hygiene. 2016;**94**(2):445-455

[21] Joseph OS, Wagura N, Jesper S. Risk perception, choice of drinking water, and water treatment evidence from Kenyan towns. Environment for Development. 2013;**2013**:1-24

[22] Mehta B, Malik M, Kumar V, Verma R, Chawla S, Sachdeva S. Knowledge attitude and practices regarding water handling and water quality assessment in a rural block of Haryana. Basic and Applied Medical Sciences. 2013;**2013**:2

[23] Nitinkumar SJ, Amaliyar. Study on knowledge, attitude and practice about purification of household water among 210 individuals of urban area of Patan District National Journal of. Community Medicine. 2019;**10**(7)

[24] Ibrahim J, Sufiyan M, Olorukooba A, Gobir A, Adam H, Amadu L. Knowledge, Attitudes, and Practices of Household Water Purification among Caregivers of Under-Five Children in Biye Community, Kaduna State. Wolters Kluwer - Medknow; 2017

Chapter 3

Reverse Osmosis in Industrial Wastewater Treatment Units

Yehia A. Shebl

Abstract

The MENA region faces a severe water crisis, prompting governments to take action by improving irrigation methods, treating and reusing sewage and agricultural wastewater, and issuing restrictions regulating industrial wastewater discharge. As a result, many large factories have established industrial wastewater treatment plants to recycle water, reduce reliance on external sources, comply with environmental regulations, and implement MLD or ZLD principles. This chapter will focus on industrial wastewater treatment using reverse osmosis (RO) membranes. It will cover the treatment of various contaminants such as nitrogen, phosphorus, COD, BOD, TOC, and heavy metals. It will discuss different treatment methods and technologies to produce reusable water while achieving MLD and ZLD principles.

Keywords: RO, IWWTP, brine desalination, ZLD, MLD, effluent environmental impacts

1. Introduction

How is water recycled in your space? Water reuse and recycling have become inevitable, especially in areas where water is scarce. Water scarcity is a global problem as most of the water on the surface of the planet is salty water, whether it is of high salinity, as in the seas, oceans, and salty lakes, or water of medium salinity, as in most of the wells waters, and that water represents about 97% of the total water present. While fresh water is mostly confined to snow in the north and south poles, which represents about 2% of the amount of water present, and the remaining 1% is divided between 0.6% fresh water in wells, and 0.3% represents moisture water in the atmosphere, and the remaining only about 0.1% of all available water resources is fresh surface water in rivers and freshwater lakes. Besides that, the easy-to-use surface freshwater is limited to about 0.1% of all water resources, and that represents a natural physical water scarcity; however, most of that water is exposed to different kinds of anthropogenic pollution, making it needs further treatment before using and leading to continued pressure on that limited water resources.

The massive industrial development in the last century and the spread of huge industrial complexes and their need for large quantities of water of different quality led to another kind of pressure on the limited water resources, in addition to the negative impact of the industrial wastewater of those industrial complexes in case it was discharged without treatment or with partial treatment to different water bodies, whether fresh or not fresh or seawater, so the safe disposal of these liquid wastes or recycling water and reusing it in various industrial processes has also become an indivisible necessity.

"*Save it before it is too late*", this slogan must have its meaning present in the minds of all those responsible for the industrial units and facilities that already exist or are under construction because one of the biggest reasons for the lack of proper and safe disposal of the industrial wastewater is the lack of interest or full knowledge of those responsible for the industry about the extent of the danger of these wastes on the environment and in the core of its water sources and its exposure to pollution. Furthermore, let us look from a narrow perspective and it will negatively affect the quality of the feed water for those factories themselves, which may lead to an increase in the cost of water treatment required for industry or affect the efficiency of the industrial process itself in an endless cycle of increasing pollution and additional treatments.

One of the best-applied methods to saving water is reusing or recycling it, where the most appropriate is to apply this at the industrial level, as the water required for industry varies in quality from one industry to another, as well as for various uses within the same industry, as (cooling water – manufacturing products – steam production – a carrier of raw materials or waste – or a solvent), and also as water reused at the industrial level that will not be affected by psychological and societal acceptance, as in the reuse of water for drinking purposes.

There are many water treatment technologies used in the treatment of industrial wastewater; probably the most prominent of them are; physical treatment like (screening, mixing, sedimentation, flotation, filtration, and gas transfer) and chemical treatment in which the removal or conversion of contaminants is carried by the addition of chemicals or by other chemical reactions like (precipitation, oxidation/reduction, neutralization, adsorption, and disinfection), also biological treatment in which the removal of contaminants is carried by biological activity to remove the biodegradable organic substances whether colloidal or dissolved and nutrients like nitrogen and phosphorus from the industrial wastewater using one or all of aerobic, anaerobic, and anoxic biological treatment methods.

Industrial wastewater treatment is a general concept, and reuse is a particular case where after applying the recommended treatment method to remove different contaminates, the question remains, is this water suitable for the type of application that will be reused through it? Possibly one of the substantial and essential treatment methods is the removal of salts through reverse osmosis (RO) membrane technology.

Despite the wide use of reverse osmosis (RO) membrane technology in the treatment of industrial wastewater for reuse, this requires several critical challenges, one of which is; due to the higher sensitivity of these membranes; they require complex primary treatment, which is considered not only every industry has its industrial wastewater case or every factory, but every stream inside the factory is evaluated as a certain case study and needs unique treatment methods that achieve the best-needed quality with the lowest costs. While the other is how to safely dispose of the resulting concentrated solution, whether by achieving the principle of minimum liquid discharge (MLD) using evaporation lakes, deep injection wells, or drainage on seawater after fulfilling the necessary environmental conditions, or thermal evaporation and crystallization achieving the principle of zero liquid discharge (ZLD), while the resulting desalinated water may not be suitable for use directly in some cases, and it needs certain additions or additional treatments before using it, depending on the type of application. On the other hand, the selection of the RO unit's proper design recovery, membrane types, flux, and configuration is another one of the most important points for sustainable RO technology application in industrial wastewater treatment and reuse.

In this chapter, we will address some important points that must be taken into consideration when designing and implementing the various stages of the treatment and reuse of industrial wastewater.

2. Pretreatment of industrial wastewater

The primary treatment of industrial wastewater is the cornerstone and depends on to what extent its efficiency could achieve the maximum benefit from that wastewater, whether by direct reuse, partially desalting using RO technology, or reaching demineralized water.

The first step is an accurate knowledge of the nature and sources of industrial wastewater based on a good knowledge of the industry processes, places of drainage, and the nature of its being continuous or patched (intermittent) streams of industrial wastewater, so you need to know the amount and frequency of each stream as well as their specifications, the next table help to know the nature of each stream flow (**Table 1**).

Also, it is essential to know the sources of raw feed water for the industry in which industrial waste treatment is to be done, as it gives an initial idea of the nature of wastewater composition, or at least the general tendency. For example, if the wastewater results from raw water from wells, it is necessary to analyze wastewater for elements such as silica, iron, manganese, calcium, magnesium, barium, and strontium. But if the raw water is surface fresh water, the the big focus will be on its organic load, silica, and so on. Also, kind of water treatment techniques used in the utility section of the industry where it is possible to maximize its recovery or rearrange it to make it more suitable and reduce the waste resulting from it, or parts of it can be used to treat wastewater or mix portion of the pretreated industrial wastewater with feed water for some of its units.

In the following, we will review some of the significant parameters of industrial wastewater, which must be closely monitored, and some of their effects on the treatment stages will be shown.

Streams		Units	Stream 1	Stream 2	Stream 3	Stream 4
Name						
Continuous Flow		m^3/h				
Peak	Flow	m^3/h				
	Duration	Hrs				
Patch (Intermittent)		Volume/ time				

Table 1.
Demonstrates a detailed approach to documenting different sources and quantities of industrial wastewater streams.

2.1 Temperature

The continuous follow-up of the change in the temperature of the industrial wastewater is a heightened concern because the shift in it may be unlike surface water or the well water is not always linked to the change in the ambient temperature during the different seasons, as there may be sources of industrial wastewater associated with a big rise in temperatures such as steam condensate drain which could be ranged from 50° C to 90° C, or some exothermic reactions which make the industrial waste stream temperature reaches 80° C. So identifying the temperature ranges of each industrial wastewater stream are very important because, based on the type of subsequent treatment of that water, the extent to which it needs to be cooled or not will be determined, and whether this stream (which may be small in quantity) can be separated and cooled separately before mixing it with the rest of the waste streams, to prevent the high temperature of the mixed (equalized) industrial wastewater above the recommended temperature for the subsequent treatment technologies like for ultrafiltration (UF) organic membrane (maximum temperature is 40° C) or RO membranes (maximum temperature is 45° C). Or there is a type of treatment that will be applied to this stream individually. Or on the contrary, blending it with the rest of the industrial wastewater sources may maintain an average temperature in different seasons, leading to improving some types of treatments that are greatly affected by the extreme drop in temperatures, such as biological treatment, coagulation processes, and also reverse osmosis, which requires certain precautions during the design phase to maintain the desired efficiency of these units.

2.2 Organic content

There are several ways to express the organic load of industrial wastewater, including the chemical oxygen diamond (COD), the biological oxygen diamond (BOD), and also the total organic carbon (TOC).

2.2.1 COD measurement

COD represents the quantity of dissolved oxygen in the water that must be present to oxidize organic materials. As a pollution measuring tool, COD is used to measure the short-term influence that wastewater effluents will have on the receiving water bodies' oxygen levels. So, COD is an essential measurement that helps detect the organic pollutant amount and follows the efficiency of different treatment techniques to ultimately limit pollution in water.

COD measuring may be essential when treated wastewater is discharged into the environment. High levels of industrial wastewater COD indicate concentrations of organics that could deplete dissolved oxygen in the water, leading to adverse environmental and regulatory significances. But when using the pretreated wastewater as feed water for RO system, TOC measuring is the best way to get the total organic loads that may be above the RO membranes manufacturers recommendations.

COD measurement has many interferences which may be present on the industrial wastewater and cannot be dependent on measurement for assets the organic loads before the RO system as interference from chloride, florid, and bromide, chromium, nitrite, sulfite ions where mercuric sulfate that eliminates chloride interference up to 2000 mg/L and samples with higher chloride concentrations are typically diluted or may be removed by precipitation with silver ion and filtration before digestion. Some aromatic compounds like Pyridine and related compounds resist oxidation, and volatile organic compounds will react in proportion to their contact with the oxidant; in contrast, straight-chain aliphatic compounds are oxidized more effectively in the presence of a silver sulfate catalyst. Also, some organic compounds are not oxidized completely with the COD method like urea compounds not properly appearing in the COD measurement, so using of COD only to measure the effectiveness of the pre-treated wastewater is not accurate, and TOC measuring is a more accurate, faster, and more sharp method for the organic content [1].

The total organic carbon (TOC) measurement is essential to assessing the pre-treatment process capabilities before the RO membrane system in industrial waste-water treatment units. TOC is a measure of the total amount of organic compounds present in water and is used as an indicator of the quality and suitability of water and can provide valuable information about the efficiency of the pretreatment process and the potential for fouling of membranes in the RO system.

TOC limits before the RO membrane system will depend on the specific design consideration of the RO system, like; RO system recovery and choosing the type of RO membrane in terms of its fouling resistance capabilities. In general, it is recommended to keep the TOC levels as low as possible to minimize the risk of fouling the RO membranes. For industrial water treatment applications, the American Membrane Technology Association (AMTA) recommends a maximum TOC concentration of 5 mg/L [2].

The presence of nutrients in the pretreated wastewater can potentially impact the permissible limits of TOC before the RO membrane system. Nutrients, such as nitrogen and phosphorus, can stimulate the growth of microorganisms, even if it is less than 5 mg/l, so it is preferred to decrease TOC concentration to less than 2 ppm in case of the presence of residual nutrition in the pretreated wastewater, or it may be required to dose nonoxidizing biocide like 2,2-dibromo-3-nitrilopropionamide (DBNPA) whether continuous or intermittent shock doses to prevent or reduce RO membrane biofouling rates.

Both TOC and COD are commonly used to measure the concentration of organic matter in water. TOC measures the total amount of organic compounds present in water, while COD measures the oxygen-depleting capacity of organic compounds in water.

In general, TOC is considered a more comprehensive measure of organic matter in water because it includes a wider range of organic compounds, including both bio/chemical degradable and nondegradable compounds. COD, on the other hand, only measures the oxygen-depleting capacity of biodegradable organic compounds.

TOC is generally considered to be more accurate than COD because it includes a wider range of organic compounds. Therefore, in terms of interferences, TOC is less subject to interference from various sources.

For this reason, TOC has generally been considered a more reliable indicator of the quality and suitability of water for various applications, including use in RO systems.

2.2.2 Biodegradability of an industrial wastewater

It is important to note that the biodegradability of industrial wastewater may vary depending on the specific contaminants present and the conditions in which the wastewater is treated. Therefore, it is important to conduct thorough testing in order to accurately determine the biodegradability of a particular industrial wastewater.

BOD5 is commonly used to measure the strength of wastewater and the effective-ness of the biological treatment processes.

There are several factors that can interfere with BOD measurements in industrial wastewater. These include:

Inorganic substances: Inorganic substances, such as sulfur, chlorine, and ammonia, can interfere with BOD measurements by reacting with the oxygen that is used in the test. This can lead to inaccurate results [1].

- pH: The pH of a sample can affect the availability of oxygen to microorganisms and can also interfere with the accuracy of BOD measurements. So, if the pH of the sample is not between 6 and 9, the pH of the sample can be adjusted to a neutral value (around pH 7) to minimize interference from changes in pH.

- Nutrient levels: The presence of nutrients, such as nitrogen and phosphorus, can affect the rate of biological activity and the accuracy of BOD measurements, where the addition of nutrients, such as nitrogen and phosphorus, can help to stimulate the growth of microorganisms and improve the accuracy of BOD measurements.

- Interference by other compounds: Other compounds, such as certain types of surfactants and detergents, may interfere with the BOD measurement by reacting with oxygen or by inhibiting the growth of microorganisms [3].

- Incomplete decomposition: If the sample is not allowed to decompose completely, the BOD measurement may be underestimated, so you may need to allow for complete decomposition to obtain accurate BOD measurements, like using BOD20.

BOD20 [4, 5] is a similar measure, but the test is conducted for 20 days rather than 5 days. This can provide a more accurate measure of biodegradability, as some substances may take longer than 5 days to break down fully. However, the BOD20 test is not as widely used as the BOD5 test, as it takes longer to conduct and requires more resources.

In general, the BOD5 test is considered sufficient for most purposes, but the BOD20 test may be used in cases where a more accurate measure of biodegradability is required or if the substance being tested is known to take longer than 5 days to break down.

There are several factors that can hinder the biological treatment of industrial wastewater. Some of these factors include:

High levels of toxins or other contaminants: If the wastewater contains high levels of toxins or other contaminants, it may be more difficult for microorganisms to break down the organic matter in the wastewater.

There are many toxins or chemical types that can hinder the biological treatment of industrial wastewater. Some examples of toxins that have been shown to have negative impacts on the biological treatment process include:

- Cellulose is a type of organic matter that is resistant to decomposition, and it can interfere with the microorganisms that are used in the BOD test to measure the amount of oxygen required to break down the organic matter in the wastewater. This can result in an underestimation of the actual BOD of the wastewater, leading to false low readings. Cellulose compounds can be degraded by biological treatment of industrial wastewater, but it may be more challenging than other organic matter types. Cellulose is a complex carbohydrate that is found in plant cell walls and is resistant to decomposition due to its complex structure and the fact that it is highly crystalline, which makes it difficult for microorganisms to access and break down. However, certain microorganisms, such as fungi and some bacteria, can break down cellulose or may need to reduce its concentrations using physical treatment methods like sedimentation, filtration, and centrifugation to remove cellulose from the source. Also, chemical treatment methods involve using chemicals to break down the cellulose, like the use of enzymes or chemicals like sodium hydroxide or hydrochloric acid [6, 7].

- Heavy metals: Heavy metals such as lead, mercury, and cadmium can be toxic to the microorganisms responsible for breaking down organic matter in wastewater. For example, a study published in the Journal of Environmental Management found that high concentrations of lead and mercury inhibited the degradation of organic matter in wastewater treatment systems [8].

- Pesticides: Pesticides such as organophosphates and carbamates can be toxic to the microorganisms responsible for breaking down organic matter in wastewater. For example, a study published in the journal Water Research found that pesticides in wastewater significantly reduced the rate of organic matter degradation [9].

- Polycyclic aromatic hydrocarbons (PAHs): PAHs are a group of chemicals formed during the incomplete burning of organic matter. They are toxic to the microorganisms responsible for breaking down organic matter in wastewater. For example, a study published in the journal Environmental Science and Technology found that high concentrations of PAHs inhibited the degradation of organic matter in wastewater treatment systems [10].

- Endocrine disrupting chemicals (EDCs): EDCs can interfere with the normal functioning of the endocrine system. They are toxic to the microorganisms responsible for breaking down organic matter in wastewater. For example, a study published in the journal Environmental Science and Technology found that EDCs in wastewater significantly reduced the rate of organic matter degradation [11].

It is important to note that these are just a few examples of toxins that can hinder the biological treatment of industrial wastewater, and there are many other types of toxins that can have similar impacts.

Lack of nutrients: Some industrial wastewater may be deficient in essential nutrients, such as nitrogen and phosphorus, which are required for the growth and activity of microorganisms.

Overall, the success of a biological treatment process for industrial wastewater will depend on various factors, including the specific contaminants present in the wastewater and the conditions in which the treatment is carried out.

2.3 Oil and grease

Oil and grease can be difficult to remove completely from industrial wastewater before the RO membranes system. It is necessary to effectively remove oil and grease from wastewater before it is treated using RO membranes to minimize their adverse impacts, where oil and grease can attach and accumulate on the surface of the RO membrane causing severe and may lead to irreversible organic fouling. RO Membrane manufacturers recommended that oil and grease concentrations should be less than 0.1 mg/l. Several techniques can be used to remove oil and grease from industrial wastewater, depending on the concentration of these contaminants. It may be necessary to use several successive methods to achieve the required quality and also according to the extent of the tendency and types of other pollutants associated with this industrial wastewater and the possibility of separating the sources of oils to treat them alone or not, especially if the concentration is high.

Gravity separation (Skimming): This involves using a floating device to physically remove oil and grease from the surface of the wastewater, where allowing the wastewater to sit in a settling tank, where the oil and grease float and can be separated by a skimmer [12].

Coagulation and flocculation: In this method, chemicals are added to the wastewater to cause the oil and grease to clump together, forming larger particles that can be more easily removed [13].

One of the important technologies that provide effective removal of many wastewaters contaminates is dissolved air flotation (DAF) is a wastewater treatment process that uses coagulation/flocculation combined with dissolved air to create tiny air bubbles that can attach to contaminants in the water, causing them to float to the surface and heavy suspended solids can sink down and then removed by a scrubber [14]. DAF is often used to remove oil and grease from industrial wastewater, as well as other suspended solids and some types of organic matter [15]. Here are a few reasons why DAF systems may be particularly important in the treatment of oil and grease in industrial wastewater:

- High removal efficiency: DAF systems are generally able to achieve high removal efficiencies for oil and grease, often in the range of 95–99%. This makes them an effective option for reducing the amount of oil and grease in wastewater to levels that meet discharge standards.

- Ability to treat a wide range of wastewater flows: DAF systems can be designed to treat a wide range of wastewater flows, from small to large volumes. This makes them a flexible option for treating oil and grease in industrial wastewater.

- Simple operation and maintenance: DAF systems are relatively simple to operate and maintain, which can reduce the overall cost of treatment.

However, DAF systems have some limitations that can affect their ability to effectively remove oil and grease from industrial wastewater. Some of these limitations include:

- High oil and grease levels: DAF systems are generally most effective at removing relatively low levels of oil and grease from wastewater. At higher concentrations, the oil and grease may not be fully removed, or the DAF system may need a pre-treatment for high oil and grease concentration using API, or cyclones techniques.

- Emulsified oil and grease: As mentioned in a previous answer, emulsified oil and grease can be more difficult to remove from wastewater than nonemulsified oil and grease. DAF systems may be less effective at removing emulsified oil and grease.

- The pH of the wastewater: The effectiveness of DAF systems can be affected by the pH of the wastewater. At high pH values (above 9), the air bubbles may not dissolve as effectively, which can reduce the overall effectiveness of the DAF system. At low pH values (below 6), the air bubbles may dissolve too quickly, which can also reduce the effectiveness of the DAF system.

Biological treatment: Certain types of bacteria can break down oil and grease into simpler, water-soluble compounds. This process can be done in a specialized bioreactor or as part of a broader wastewater treatment process. High concentrations of oil and grease can also interfere with the biological processes used to treat the wastewater, which can reduce the overall performance of the treatment plant [16].

Absorption: This involves using a solid material, such as clay or a synthetic polymer, to absorb the oil and grease from the wastewater.

Carbon filters can be used to treat industrial wastewater to remove residual traces of oil and grease, and other contaminants such as refractory organic compounds. Carbon filters work by adsorbing contaminants onto the surface of the carbon, then after accumulating on carbon media it can then be removed by backwashing and rinsing or either regenerating or replacing the carbon media. Carbon filters are typically used as a final step in the treatment of industrial wastewater to remove any remaining contaminants that other treatment methods may not have effectively removed or to reach out to a high-quality effluent. In addition, they are often used in combination with other treatment technologies, such as physical separation, chemical treatment, and biological treatment, to provide a high level of contaminant removal [17].

One of the main advantages of using carbon filters for oil and grease removal is their ability to effectively remove a wide range of contaminants. Carbon filters can also be effective at removing contaminants that have a low solubility in water, which can make them difficult to remove using other methods. However, carbon filters have some limitations that should be considered when using them for oil and grease removal. For example, carbon filters may become saturated with contaminants over time, which can reduce their effectiveness and increase the frequency of filter changes.

Membrane filtration: This method involves passing the wastewater through a membrane with small pores that can remove oil and grease by size exclusion. But also, it is important to effectively remove oil and grease from wastewater before it is treated using membranes like ultrafiltration (UF) to minimize these negative impacts, and it is considered a final removal step of only residual concentrations [18].

It is important to carefully consider the type and volume of oil and grease in the wastewater and the desired treatment level when selecting a treatment method.

One of the important contaminants that hurt the removal efficiency of oil and grease from industrial wastewater is the presence of emulsified agents in industrial wastewater which can make the removal of oil and grease more challenging and reduce the efficiency of treatment processes. Emulsified agents can cause oil and water to mix and form an emulsion, a stable mixture of oil droplets suspended in water [19].

Some common emulsified agents that may be present in industrial wastewater include surfactants, soaps, and detergents. These agents can interfere with the ability of physical separation methods, such as skimming, floating, and sedimentation, to effectively remove oil and grease from the wastewater. They can also make it more difficult to use chemical methods, such as coagulation and flocculation, to remove oil and grease by forming stable emulsions that are resistant to flocculation.

Biological methods, such as biodegradation, can also be affected by the presence of emulsified agents. Some microorganisms may be able to break down the emulsified agents, but this can also consume some of the oxygen in the wastewater, which can be detrimental to the overall treatment process.

Membrane filtration may be less affected by the presence of emulsified agents, as it relies on size exclusion rather than chemical or biological processes to remove contaminants. However, the efficiency of membrane filtration can still be reduced if the emulsified agents coat the membrane or if they form stable emulsions that are too small to be effectively removed by the membrane.

There are some considerations that should be taken in the design and operations of the oil and grease removal process, and the following are the most common two of them;

- De-emulsification is the process of breaking down an emulsion, or a mixture of two immiscible liquids, such as oil and water, into their components. In wastewater treatment, de-emulsification is often used to separate oil and grease from the water so that the water can be treated and returned. The chemical methods for de-emulsifying oil and grease in wastewater include mechanical separation and chemical separation;

- Mechanical separation involves physically separating the oil and grease from the water. This can be done using devices such as centrifuges.

- Chemical separation involves adding chemicals to the wastewater to help break down the emulsion. Common chemicals used for this purpose include surfactants, molecules that can help destabilize the emulsion, and emulsion breakers, encouraging the oil and water to separate.

- Saponification can occur in industrial wastewater treatment as part of the process of removing oil and grease from the water. The saponification process involves reacting an alkali, such as sodium hydroxide, with a fat or oil to produce soap. This can lead to solubilizing the oil and grease, making it difficult to separate from the water. Several factors can affect the occurrence of saponification in wastewater treatment, including the pH of the wastewater and the type of oil and grease present, where some oils and greases are more easily saponified than others due to their chemical composition. For example, animal fats and vegetable oils are more easily saponified than mineral oils. In general, higher pH values and the presence of more easily saponifiable oils and greases will tend to promote the saponification process.

2.4 Heavy metals occurrence and treatment

Heavy metals are naturally occurring elements with a high atomic weight and density at least five times greater than that of water. They are commonly found in industrial wastewater, and their presence can negatively impact the environment and human health. Also, have a negative impact on the performance of RO membranes, which are commonly used to treat and reuse industrial wastewater. When present in high concentrations, heavy metals can foul RO membranes. Fouling of RO membranes by heavy metals can occur through a variety of mechanisms, including adsorption of the metal ions onto the membrane surface, precipitation of the metal ions within the membrane pores, and the formation of metal hydroxides or metal oxides or insoluble metal compounds on the membrane surface. In addition to fouling, heavy metals in the feed water can also lead to toxic disinfection byproducts, which can occur when the heavy metals react with disinfectants (such as chlorine) used to pretreat the wastewater, resulting in the formation of harmful compounds.

Heavy metals are commonly found in industrial wastewater due to their use in various industrial processes. Some examples of heavy metals that may be present in industrial wastewater include:

- Lead: Lead is often used to produce batteries, pigments, and metal alloys. It can be found in industrial wastewater due to its use in these processes and from the corrosion of lead pipes.

- Mercury: Mercury is used to producing chemicals, pesticides, and pharmaceuticals. It can also be found in industrial wastewater due to its use in the extraction of gold and silver from ore.

- Chromium: Chromium is used in the production of stainless steel, leather tanning, and wood preserving. It can be found in industrial wastewater due to its use in these processes.

- Cadmium: Cadmium is used in the production of batteries, pigments, and coatings. It can be found in industrial wastewater due to its use in these processes and from the corrosion of cadmium pipes.

- Arsenic: Arsenic is used in the production of pesticides and herbicides. It can also be found in industrial wastewater due to its presence as a natural contaminant in some ore deposits.

It is important to properly treat industrial wastewater containing heavy metals before it is processed using RO membranes (mostly should be less than 0.05 mg/l) to minimize the negative impact on the membrane's performance [20]. This may involve using physical, chemical, or biological treatment methods to remove or neutralize the heavy metals in the wastewater.

There are several important considerations to take into account when removing heavy metals from industrial wastewater:

- Type and concentration of heavy metals: The specific type and concentration of heavy metals present in the wastewater will determine the most appropriate treatment method. Some treatment methods are more effective at removing certain heavy metals than others, so it is important to carefully analyze the wastewater to determine the most suitable approach.

- Environmental impact: Disposal of heavy metal-containing waste, it is important to properly dispose of any waste generated during the treatment process that contains heavy metals. This may involve safely storing the waste in a secure location or properly disposing of it through a licensed waste management facility. Also, treatment methods may have negative impacts on the environment, such as by producing harmful byproducts or consuming large amounts of energy.

- Cost and feasibility: The cost and feasibility of the treatment method should be considered when selecting the most appropriate approach. Some treatment methods may be more expensive or logistically challenging to implement, so it is important to carefully weigh the costs and benefits of each option.

- Effectiveness: The effectiveness of the treatment method in removing the heavy metals from the wastewater should be a top consideration. It is important to select a treatment method that can effectively reduce the concentration of heavy metals in wastewater to acceptable levels.

- Health and safety: The health and safety of workers involved in the treatment process should be a top priority. It is important to ensure appropriate safety measures are in place to protect workers from exposure to hazardous materials.

There are a variety of treatments that can be used to remove heavy metals from water and other materials. These include physical, chemical, and biological treatments.

One example of a physical treatment for heavy metal removal is sedimentation, in which the heavy metals are separated from the wastewater by settling. Heavy metals can precipitate from wastewater by raising the pH to a level above their respective precipitation pH. Precipitation pH is the pH at which a particular metal ion will begin to precipitate out of the solution as a solid. Different metal ions have different precipitation pH values, so the specific pH required to precipitate a particular metal will depend on the type of metal. For example, if the wastewater contains lead ions and the pH is raised to 9.5, the lead ions may begin to precipitate out of the solution as a solid. It is important to note that simply raising the pH of the wastewater may not be sufficient to completely remove all heavy metals. Other treatment methods, such as chemical precipitation or ion exchange, may be necessary to effectively remove the heavy metals from the wastewater [21, 22].

Chemical treatments for heavy metal removal include the use of chelating agents, which can bind to the heavy metals and allow them to be removed from the water.

Biological treatments for heavy metal removal include the use of bacteria that can absorb and remove heavy metals from the water.

Once the heavy metals have been removed from the water, they can often be recovered and reused. This can be done through metal recovery, which involves separating the heavy metals from the material they were removed from and purifying them for reuse. By using this treatment process, the factory can reduce its environmental impact and recover valuable resources for reuse [23].

Several methods can be used to remove heavy metals from industrial wastewater. Here are some examples of treatment methods that can be used to remove specific heavy metals from wastewater:

- Lead: Chemical precipitation, ion exchange, and electrocoagulation are all effective methods for removing lead from industrial wastewater.

- Mercury: Activated carbon adsorption and chemical precipitation effectively remove mercury from industrial wastewater.

- Chromium: Chemical precipitation, ion exchange, and electrocoagulation are all effective methods for removing chromium from industrial wastewater.

- Cadmium: Chemical precipitation, ion exchange, and electrocoagulation are all effective methods for removing cadmium from industrial wastewater.

- Arsenic: Chemical precipitation, ion exchange, and coagulation/flocculation are all effective methods for removing arsenic from industrial wastewater.

2.5 Hardness removal

Hardness in industrial wastewater is often caused by high concentrations of calcium, magnesium, carbonate, and sulfate ions, which can come from various industrial processes. These ions can cause various problems, including scale formation in pipes and equipment, and can interfere with the effectiveness of wastewater treatment and reuse in certain industrial processes. Softening industrial wastewater before it is treated with RO can help to increase the operated recovery of the RO unit with less dosage of antiscalant, so increasing the wastewater reused.

Hardness in industrial wastewater can come from a variety of sources, including:

- Industrial processes: Many industrial processes, such as power generation, oil and gas production, and metal finishing, can produce wastewater with high levels of hardness.

- Cooling water: Water used for cooling industrial equipment, such as cooling towers blowdown, can become hard due to the accumulation of calcium and magnesium ions depending on the cycle of concentration (COC).

- Boiler water: Boiler blowdown involves the removal of hard water from a boiler to maintain specific parameters within certain limits and prevent issues such as corrosion, scale, and carryover. It is also used to remove suspended solids that may be present in the system.

- Groundwater: In some cases, the source of raw feed water used in industrial processes may have a naturally high hardness level, like the groundwater.

Several methods can be used to remove hardness from industrial wastewater, including:

- Chemical treatment: This involves adding chemicals such as lime or soda ash to the wastewater, which can react with the calcium and magnesium ions to form a precipitate that can be separated from the water.

- Ion exchange: This involves passing the wastewater through a bed of resin beads that are charged with sodium ions. The calcium and magnesium ions in the water are attracted to the resin beads and exchange places with the sodium ions, leaving the water with a lower hardness level.

- Nanofiltration: This filtration process uses a membrane to remove ions and other contaminants from the water. It effectively removes hardness but requires a comprehensive pretreatment like RO membranes.

- Electrodialysis: This process uses an electric current to separate ions in the water based on their charge. It effectively removes hardness but is expensive and requires specialized equipment.

It is important to choose the most appropriate method for removing hardness from industrial wastewater based on the specific needs and constraints of the application.

Some important considerations that should be taken into account when deciding lime soda softening and caustic soda softening for the treatment of industrial wastewater:

Cooling tower blowdown water contains a dispersant and antiscalant that hindered or disturb the coagulation-precipitation process. Where dispersants are chemicals added to the cooling water to prevent the formation of scale and the precipitation of solids, they work by inhibiting the aggregation of particles and keeping them suspended in the water. This can make it difficult or impossible to remove contaminants through coagulation-precipitation. Also, antiscalants are chemicals added to the cooling water to prevent scale formation on surfaces. They work by inhibiting the precipitation of minerals such as calcium and magnesium. However, these chemicals can interfere with the coagulation-precipitation process by preventing the formation of the necessary flocs or aggregates of particles that are necessary for the process to be effective. So it is important to degrade this chemical and inhibit its functions by using a strong oxidizing agent like chlorine with a sufficient concentration and contact time to effectively inactivate those chemicals before the coagulation-precipitation process.

Ferric chloride and alum is a commonly used coagulants in the treatment of industrial wastewater. However, there is a negative impact that should be considered when using ferric chloride as a coagulant before the RO membrane system especially at softening clarifiers where pH is high enough to dissolve part of those minerals: membrane fouling: where ferric chloride can cause fouling of the RO membrane, which can reduce its efficiency and require more frequent cleaning.

It is important to consider these negative impacts carefully when deciding whether to use ferric chloride or alum as a coagulant before an RO system.

In the chemical softening process of wastewater treatment, the pH of the water is typically raised during the softening process and then lowered during the neutralization step. The alkalinity of the wastewater can play a role in the amount of chemicals (lime soda or caustic soda) required to raise the pH to the desired level for softening. Alkalinity is a measure of the water's ability to neutralize acids and is typically expressed in terms of the concentration of bicarbonate, carbonate, and hydroxide ions in the water.

If the alkalinity of the wastewater is high, it may take more chemicals to raise the pH to the desired level for softening. This is because the water's high alkalinity indicates the presence of a large amount of bicarbonate, carbonate, and hydroxide ions, which can neutralize the alkaline chemical added to raise the pH. As a result, more alkaline chemicals may be required to overcome the buffering effect of the alkalinity and achieve the desired pH.

On the other hand, if the alkalinity of the wastewater is low, it may take less chemicals to raise the pH to the desired level for softening. In this case, fewer bicarbonate, carbonate, and hydroxide ions are present to neutralize the alkaline chemical, so fewer chemicals are required to achieve the desired pH.

After the softening process, the pH of the wastewater is typically lowered during the neutralization step. To neutralize the excess lime or caustic soda, an acid such as sulfuric acid or hydrochloric acid is added to the water. The acid reacts to effectively neutralize the pH of the water. It is important to carefully control the amount of acid added to the water, as adding too much acid can result in a pH that is too low, which can have negative effects on the environment or downstream processes.

The buffering effect in wastewater is a result of the presence of ions that can neutralize acids or bases, preventing significant changes in pH. High concentrations of ions such as ammonia/nitrate or bicarbonate can contribute to the buffering effect in wastewater.

Ammonia (NH_3) and nitrate (NO3-) ions can act as weak bases in water, neutralizing acids and helping to maintain a relatively stable pH. Bicarbonate (HCO3-) ions can also act as a buffer in water, neutralizing both acids and bases and helping to maintain a relatively stable pH.

The buffering capacity of wastewater can have an impact on the effectiveness of pH adjustment or neutralization processes. If the wastewater has a high buffering capacity, it may take more acid or alkaline chemicals to achieve the desired pH change. Conversely, if the wastewater has a low buffering capacity, it may take less acid or alkaline chemical to achieve the desired pH change.

It is important to carefully monitor the pH and buffering capacity of wastewater during treatment processes and adjust the chemical dosage as needed to effectively adjust the pH or neutralize the water.

The addition of lime or caustic soda to wastewater during the softening process will typically result in an increase in the total dissolved solids (TDS) of the water. This is because the lime or caustic soda reacts with the hardness-causing ions in the water to form solid precipitates, which contribute to the TDS of the water. The neutralization step, in which acid is added to the water to neutralize the excess lime or caustic soda, will not typically result in a significant change in the TDS of the water.

It is important to note that the TDS of the water can also be affected by other factors, such as the presence of other dissolved solids in the water, the volume of water treated, and the efficiency of any downstream treatment processes. In general, it is desirable to keep the TDS of wastewater as low as possible, as high TDS can have negative effects on the environment and on any downstream processes.

2.6 Total dissolved solids

It is important to accurately determine the final TDS and other component ions of pretreated wastewater before it is treated with a reverse osmosis (RO) system. This information is critical for properly designing the RO unit and optimizing its performance.

During the pretreatment process, various chemicals may be added to the wastewater to remove contaminants or adjust the water's properties. For example, chemical precipitation or softening may be used to remove hardness-causing ions, and pH neutralization may be used to adjust the pH of the water. However, these processes can increase the TDS of the wastewater, as the chemicals added can contribute to the dissolved solids content of the water.

Accurately measuring the TDS and other component ions of the pretreated wastewater is important because it allows the RO system to be properly sized and configured to meet the specific treatment needs of the water. It also helps to ensure that the RO system is operating at optimal efficiency and can effectively remove contaminants from the water.

It is generally recommended to measure the TDS and other component ions of the wastewater at various points throughout the treatment process, to gain a comprehensive understanding of the water's quality and to make any necessary adjustments to the treatment process.

Also, it is important to consider the potential for changes in the characteristics of the wastewater during the design and operation of a reverse osmosis (RO) system. For example, in situations where different streams of wastewater are combined and neutralized, there is a risk that changes in the flow or concentration of one of the streams could affect the final TDS and other ion concentrations of the RO feed water.

To address this risk, it is important to carefully design the RO system to be able to handle the worst-case scenario. This may involve selecting RO membranes that are resistant to fouling and able to handle a wide range of water qualities, selecting a high-pressure pump that is capable of operating effectively under varying conditions, and specifying piping materials that are compatible with the wastewater being treated.

By designing the RO system to be able to handle the worst-case scenario, it is possible to ensure that the system is sustainable and durable and able to effectively treat the wastewater even if there are changes in the flow or concentration of the different streams. It is also important to regularly monitor the water quality and adjust the treatment process as needed to maintain the efficiency and effectiveness of the RO system.

3. Conclusions

Industrial wastewater treatment and reuse are gaining high importance in many parts of the world. This chapter gives an overview of the considerations involved in the treatment and reuse of industrial wastewater, with a focus on the pretreatment of wastewater before the reverse osmosis (RO) treatment. It highlights the various factors that have contributed to the growing importance of industrial wastewater treatment and reuse, including environmental regulations, limited water resources, cost savings, and sustainability. The conclusion also describes the various pretreatment technologies and techniques that can be used to prepare industrial wastewater for RO treatment, such as chemical precipitation, softening, pH adjustment, oil and grease removal, biological processes, and filtration. It emphasizes the importance of properly designing and operating the pretreatment system, as well as accurately measuring the TDS and other component ions of the wastewater, to ensure the efficiency and effectiveness of the RO system. Finally, the conclusion notes that there is no one-size-fits-all treatment scheme for industrial wastewater and that the specific pretreatment steps required will depend on the characteristics of the wastewater and the specific requirements of the RO system being used.

Author details

Yehia A. Shebl
Chemistry Department, Faculty of Science, Cairo University, Giza, Egypt

*Address all correspondence to: yehia_aly@yahoo.com

References

[1] Rice EW, Bridgewater L, American Public Health Association, editors. Standard methods for the examination of water and wastewater. Washington, DC: American Public Health Association; Feb 2012

[2] American Membrane Technology Association (AMTA). Industrial Applications of Membranes. Stuart: AMTA

[3] Jones DL, Freeman C, Sánchez-Rodríguez AR. Waste water treatment. In: Encyclopedia of Applied Plant Sciences: Crop Systems. 2nd ed. Elsevier; 2016. pp. 352-362. DOI: 10.1016/B978-0-12-394807-6.00019-8

[4] Rice EW, Baird RB, Eaton AD. Standard Methods for the Examination of Water and Wastewater. 23rd ed. Washington DC: American Public Health Association (APHA), American Water Works Association (AWWA) and Water Environment Federation (WEF); 2017

[5] Von Sperling M. Wastewater characteristics, treatment and disposal. IWA Publishing; 2007

[6] Chaparro TR, Pires EC. Anaerobic treatment of cellulose bleach plant wastewater: Chlorinated organics and genotoxicity removal. Brazilian Journal of Chemical Engineering. 2011;**28**:625-638

[7] Pandey JK, Saini DR, Ahn SH. Degradation of cellulose-based polymer composites. In: Cellulose Fibers: Bio- and Nano-Polymer Composites. Berlin, Heidelberg: Springer; 2011. pp. 507-517

[8] Zhang L, Lin X, Wang J, Jiang F, Wei L, Chen G, et al. Effects of lead and mercury on sulfate-reducing bacterial activity in a biological process for flue gas desulfurization wastewater treatment. Scientific Reports. 2016;**6**(1):1

[9] Bocquené G, Galgani F. Biological effects of contaminants: Cholinesterase inhibition by organophosphate and carbamate compounds [Internet]. ICES Techniques in Marine Environmental Science (TIMES); 1998 [cited 2023 Mar 17]. Available from: https://ices-library.figshare.com/articles/report/Biological_effects_of_contaminants_Cholinesterase_inhibitation_by_organophosphate_and_carbamate_compounds/18626807/2

[10] Haritash AK, Kaushik CP. Biodegradation aspects of polycyclic aromatic hydrocarbons (PAHs): A review. Journal of Hazardous Materials. 2009;**169**(1-3):1-5

[11] Chen Y, Yang J, Yao B, Zhi D, Luo L, Zhou Y. Endocrine disrupting chemicals in the environment: Environmental sources, biological effects, remediation techniques, and perspective. Environmental Pollution. 2022;**310**:119918

[12] Edwards JD. Industrial wastewater treatment. Taylor & Francis Eboo [Internet]. Taylor & Francis; 2019 [cited 2023 Mar 17]. Available from: https://www.taylorfrancis.com/books/mono/10.1201/9781351073509/industrial-wastewater-treatment-joseph-edwards

[13] Abd El-Gawad HS. Oil and grease removal from industrial wastewater using new utility approach. Advances in Environmental Chemistry. 2014;**2014**:1-6

[14] Daud Z, Awang H, Nasir N, Ridzuan MB, Ahmad Z. Suspended solid, color, COD and oil and grease

removal from biodiesel wastewater by coagulation and flocculation processes. Procedia-Social and Behavioral Sciences. 2015;**195**:2407-2411

[15] Rocha e Silva FC, Rocha e Silva NM, Luna JM, Rufino RD, Santos VA, Sarubbo LA. Dissolved air flotation combined to biosurfactants: A clean and efficient alternative to treat industrial oily water. Reviews in Environmental Science and Bio/Technology. 2018;**17**(4):591-602

[16] Sanghamitra P, Mazumder D, Mukherjee S. Treatment of wastewater containing oil and grease by biological method - A review. Journal of Environmental Science and Health, Part A. 2021;**56**(4):394-412

[17] Fulazzaky MA, Omar R. Removal of oil and grease contamination from stream water using the granular activated carbon block filter. Clean Technologies and Environmental Policy. 2012;**14**(5):965-971

[18] Zhang L, Cheng L, Wu H, Yoshioka T, Matsuyama H. One-step fabrication of robust and anti-oil-fouling aliphatic polyketone composite membranes for sustainable and efficient filtration of oil-in-water emulsions. Journal of Materials Chemistry A. 2018;**6**(47):24641-24650

[19] Smith D, Williams F, Moffatt S. Wastewater treatment methods. In: Essential Readings in Light Metals. Champions: Springer; 2016. pp. 685-690

[20] Wenten IG. Reverse osmosis applications: Prospect and challenges. Desalination. 2016;**391**:112-125

[21] Pang FM, Teng SP, Teng TT, Omar AM. Heavy metals removal by hydroxide precipitation and coagulation-flocculation methods from aqueous solutions. Water Quality Research Journal. 2009;**44**(2):174-182

[22] Shrestha R, Ban S, Devkota S, Sharma S, Joshi R, Tiwari AP, et al. Technological trends in heavy metals removal from industrial wastewater: A review. Journal of Environmental Chemical Engineering. 2021;**9**(4):105688

[23] Selvi A, Rajasekar A, Theerthagiri J, Ananthaselvam A, Sathishkumar K, Madhavan J, et al. Integrated remediation processes toward heavy metal removal/recovery from various environments - A review. Frontiers in Environmental Science. 2019;**7**:66

Section 2

Technologies

Chapter 4

The Connection between the Impacts of Desalination and the Surrounding Environment

Adel Hussein Abouzied

Abstract

The background of water desalination is covered in this chapter, along with an analysis of the environmental issues the desalination industry faces and suggestions for how to address them, to close the gap between the growing demand for water for all purposes and the natural water resources' finite availability since the early 1970s. While a few number plants established in desert locations desalinate brackish and saline groundwater, most plants built in coastal areas desalinate seawater. Desalination of water has detrimental effects on both marine and terrestrial habitats. Desalination plants also deal with issues such as corrosion, sedimentation, membrane fouling, and scale formation, the disposal of rejected brine from coastal or inland desalination facilities and its harmful impacts on the ecosystems of the marine environment and groundwater. Focus should be placed on achieving zero-brine discharge, incorporating solar-pond technology, using renewable energy sources in desalination, and supporting research and development in the field of water desalination in order to reduce the negative effects of the desalination industry on the nation. Desalination still has difficulties in managing its waste products and minimizing its energy requirements in order to avoid negative environmental effects.

Keywords: desalination, environment, water, brine management, desalination alternatives

1. Introduction

The need for water is always rising to keep up with the demands of population growth, improving living conditions, expanding green space, rising per capita consumption, urban development, and industrial growth. According to MOEW (2010), demand management may involve increasing public awareness, water tariffs, reducing water losses, and effective water billing and bill collecting [1]. Over-pumping groundwater to meet the rising water demand has resulted in the depletion of significant aquifers and the worsening of groundwater quality [2]. The use of unconventional water sources, like treated waste-water and desalinated water, has greatly increased to close the gap between water supply and demand. However, this strategy has detrimental effects on the environment and raises the energy demand necessary

No.	Country	Total capacity (million M^3/d)	Market share (%)
1	Saudi Arabia	9.9	16.5
2	USA	8.4	14.0
3	UAE	7.5	12.5
4	Spain	5.3	8.9
5	Kuwait	2.5	4.2
6	China	2.4	4.0
7	Japan	1.6	2.6
8	Qatar	1.4	2.4
9	Algeria	1.4	2.3
10	Australia	1.2	2.0

Table 1.
Top 10 nations that use desalination, taken from Nair and Kumar.

for the expansion of water desalination, MOEW (2015). Communities have resorted to alternate water sources, such as desalination, water recycling, and water import in several places across the world where local water basins are becoming depleted [3, 4]. Desalination is the process of purifying saltwater by removing extra salt and other dissolved compounds. The World Health Organization's 500 ppm drinking water limit is reached or exceeded by this method, which lowers salt content [5, 6].

Water is a national resource that requires a national strategy for integrated water-resources management and to address the difficulties brought on by the increased water demand [7]. Desalination has solidified its place in recent years as a means of reducing the world's water shortage. Reverse osmosis is the most widely used technique in the global desalination market and is regarded as the most optimal membrane-based desalination process, producing around 50% of the desalination water. Desalination uses a lot of energy, though, and historically has relied on fossil fuel-based processes [8–10].

Since then, the desalination of brackish water and seawater has spread fast throughout the world. More than 17,000 desalination units were operational in 2013, delivering around $80x10^6$ m^3/d to 300 million people across 150 nations. By 2015, the production capacity had nearly reached $97.5x10^6$ m^3/d, and by 2050, it is anticipated that there will be $192x10^6$ m^3/d of desalinated water available [5]. The top 10 countries using desalination are listed in **Table 1**.

2. Technologies for desalination

Reverse osmosis (RO), Forward osmosis (FO), multi-effect distillation (MED), multi-stage flash distillation (MSF), Vapor-compression (VC), Ion exchange, Membrane processes, Electro-dialysis (ED), Capacitive Deionization (CDI), Nano-filtration (NF), Membrane distillation (MD), Hydration (HY), Secondary Refrigerant Freezing (SRF), Solar Still Distillation (SSD), and Solar Chimney (SC) are all processes used in water desalination plants. Many cogeneration facilities where the thermal energy needed to desalinate water are also used to generate electricity. The most common method for pumping brackish water through membranes while utilizing electrical energy is reverse osmosis (**Figure 1**) [10, 11].

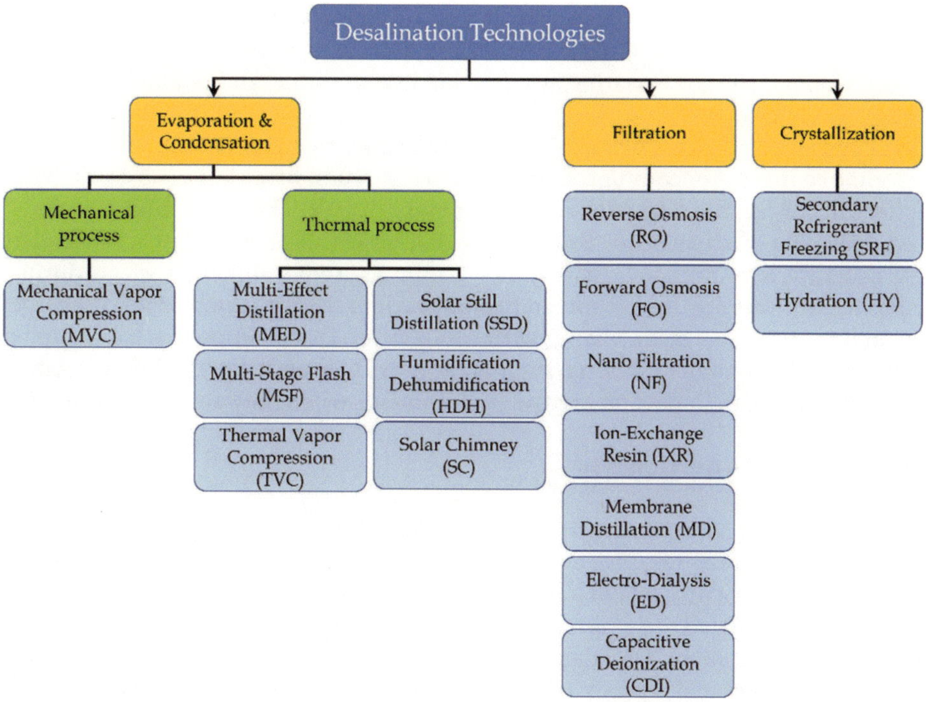

Figure 1.
Shows the grouping of desalination technologies according to their operating principles.

2.1 Environmental difficulties

The World Wide Fund (WWF) of the Global Freshwater Program criticized saltwater desalination as an expensive, energy-intensive, and method of producing drinking water that emits greenhouse gases [12]. In addition to reducing places for fishing, swimming, and enjoyment, desalination plants also release brine, which contributes to visual pollution. The produced water from the desalination plants must also undergo post-treatment, which includes treatments for organics, hypoxia, carbon dioxide (CO_2), copper (Cu), hydrogen sulfide (H_2S), hydrogen ion concentration (pH), coagulants, chlorine (Cl), and copper. Desalination plants also deal with issues like corrosion, sedimentation, membrane fouling, and scale formation. Concerns were raised by Al Asam and Rizk (2009) over the disposal of reject brine from coastal and inland desalination facilities because of their harmful effects on the ecosystems of marine environment and groundwater [13].

Most desalination processes require a lot of energy. If renewable energy sources are not employed for the production of freshwater, desalination has the potential to increase reliance on fossil fuels, raise greenhouse gas emissions, and exacerbate climate change. Surface water intakes for desalination pose a serious threat to marine life [6]. When mature fish, larvae, and other marine life are stuck in or sucked into open sea surface intake pipes, serious harm or death may result. According to the State Water Resources Control Board, the open ocean intakes utilized by California's coastal power facilities destroy 70 billion fish larvae and other marine species every year. The utilization of these open ocean intakes is being considered for desalination facilities all around California. Due to the dangerously high concentration of salts and other minerals included in brine waste, it may also represent a hazard to marine life and water quality. Because its high salinity and density, brine waste can collect in and around disposal sites, suffocating animals that live on the ocean floor and drastically changing coastal ecosystems [5, 6].

A hyper-saline slurry known as brine is created when minerals, extracted salts, and some source water combine. Compared to salt water, brine has a substantially higher salt concentration, which makes disposal difficult. The ocean is frequently used to dispose of waste brine. Brine can be discharged through diffusers or blended with other water sources to lessen salinity to minimize environmental effects during disposal. Diffusers are used to spread brine at various desalination facility discharge sites and to encourage brine mixing with ocean water. Desalination also has a number of negative environmental effects, such as excessive CO_2 emissions and waste compounds that have an impact on marine environments when they are released [6, 14].

2.2 Alternatives to desalination

Desalination is an expensive method to increase local water supply because it uses a lot of energy and has negative environmental effects. Is the average cost of ocean-water desalination a problem or a solution? Desalination plant in the Canary Islands' Lanzarote. Desalinated water is frequently 2–4 times more expensive per acre-foot than other water sources. Desalination by the ocean is ineffective. For every gallon of freshwater generated, approximately two gallons of ocean water are needed. This means that a single, massive desalination plant cannot address the issues with the local water supply. Increased regional water supplies can be achieved by water conservation, water use efficiency, storm water capture, reuse, and recycling, which are frequently more affordable than desalination. These alternatives also offer benefits that are frequently disregarded in cost–benefit analyses, such as flood control, habitat restoration, and pollution abatement [14].

2.3 Physical effects of desalination

The fundamental physical problem that water desalination presents to the environment is the temperature difference between rejected brine and feed water. The temperature of ambient saltwater is often 10 to 15 degrees Celsius lower than that of rejected brines, which is harmful to marine ecosystems. The brines released into the marine environment float on the water's surface due to their greater temperature. The water-dissolved oxygen decreases with increasing temperature, and this decline in levels of water-soluble oxygen can cause toxicity affecting marine life. The temperature variation is a minor problem as indicated by Younos (2005). This region is naturally hot, and large annual variations in temperature represent a natural phenomenon. However, persistent long-term variations in temperature of seawater can be extremely harmful and cause the death of many marine species [14].

2.4 Chemical implications of desalination

Chemicals are introduced as antiscalants during the pretreatment and chlorination procedures in the desalination business. The compounds that remain in the rejected brine are thought to be responsible for the chemical consequences of water desalination. Desalination plants in the Arabian Gulf region pump tones of metals, chemicals, and chloride into the sea each day to desalt more than 24 million m3 of seawater.

Chemical	Symbol	Use
Sodium hypochlorite or Chlorine	NaOCl or Cl	Chlorination and prevention of biological growth in membrane facilities
Ferric chloride, or Aluminum chloride	$FeCl_3$ or $AlCl_3$	Disinfectants for flocculation and removal of suspended matter
Sulfuric acid or Hydrochloric acid	H_2SO_4 or HCl	Adjustment the pH of the seawater
Sodium hexa meta-phosphate	$(NaPO_3)_6$	Prevention of scale formation
Sodium bisulphate	$NaHSO_3$	Neutralization of chlorine in the feed water
EDTA	$C_{10}H_{16}N_2O_8$	Removal the carbonate deposits
Citric acid or Sodium polyphosphate	$C_6H_8O_7$ or $NaPO_3$	Cleaning membranes three to four times

Taken from Younos (2005).

Table 2.
Chemicals used in desalination plants' pre-treatment.

Al Barwani and Purnama (2008) claim that the UAE, Saudi Arabia, Qatar, Kuwait, Bahrain, and Iran each have over 120 desalination plants that discharge daily amounts of ammonia (NH_3), 24 tons of chlorine (Cl), 65 tons of antiscalants, 300 kilograms of copper, and 65 tons of antiscalants into the Arabian Gulf. If low-quality stainless steel is utilized in the construction of desalination facilities, brine discharge will contain high quantities of iron (Fe), chromium (Cr), nickel (Ni), and molybdenum (Mo). **Table 2** provides a list of the typical pre-treatment chemicals used in desalination plants [15].

The most often used anti-fouling is the chloride (Cl) additive, according to Höpner and Lattemann (2002). Many chlorinated and halogenated organic byproducts are created when it combines with the organic molecules in seawater. Numerous studies have revealed the carcinogenicity of these substances as well as their other negative effects on aquatic life. Only 20 μg/L of Cl can significantly limit the photosynthetic of plankton. Cl level of 50 μg/L can alter the biodiversity of marine life and significantly diminish it. Lethal Cl concentrations for various fish species range from 20 to several hundred μg/L [16].

Eutrophication issues affect the desalination plant exits where polyphosphates are used. Antiscalants have a moderate to low degree of biodegradability and unidentified adverse effects. Antiscalants have an impact on the natural processes in the marine environment that include divalent metal ions, such as magnesium ions (Mg^{2+}) and calcium ions (Ca^{2+}). Copper (Cu) compounds are poisonous and inhibit the growth and reproduction of marine species in greater quantities, according to Höpner and Lattemann (2002). Cu compounds travel through the water and gather in sediments where they are ingested by benthic marine animals and enter the food chain [16].

Backwash water with coagulants and suspended debris is untreated and released into the marine environment at discharge sites. Benthic creatures are buried in the discharge sites as a result of this process, which also intensifies coloring, reduces light penetration, and raises turbidity. If discharged to surface water without treatment, the cleaning solutions and their additives, as well as the acidic (pH 2–3) and alkaline (pH 11–12) solutions are detrimental to aquatic marine life.

2.5 Biological consequences of desalination

The early loss of species in the intake zones of desalination plants was attributed by Al Dousari (2009) to the impacts of entrainment and impingement as well as the chlorination process. High biochemical oxygen demand (BOD), which causes low dissolved oxygen (DO) in saltwater, is present in the release areas of rejected brine. Rejected brines' salinity is higher than the natural ocean salinity, which has an impact on both creatures living organisms in open water and on the bottom [17, 18]. Although certain species have evolved to the natural salinity changes, the bulk of the neighboring marine animals are at risk of death due to high salinity at the discharge area of RO plants. The habitats of mangroves and the development of corals and sea grass are significantly impacted by the decrease in saltwater quality and rise in salinity levels in the vicinity of desalination facilities [18].

The phenomenon known as "brine underflows," where layers of hyper-saline solution covered the seafloor, can result from the direct discharge of brine into seas [19]. At the point of discharge, the brine concentrate is mixed as much as is practical, although this blended product frequently still tends to sink to the ocean floor. Brine underflows gradually reduce the amount of dissolved oxygen (DO) in the ocean [20]. The habitat deterioration caused by the high salinity and low DO levels, especially for benthic (bottom-dwelling) animals, can result in fewer benthic bacteria, phytoplankton, invertebrates, and fish communities. Additionally, harmful compounds that are not usually eliminated during later steps may be included in the chemicals added for the pretreatment of feed water, such as coagulants and antiscalants [21].

It is indeed possible for the number of contaminants in the brine to be 4–10 times higher than in the source water, including nitrate, arsenic, and naturally occurring radioactive elements. As a result, the direct release of brine into marine and coastal waterways has the potential to degrade water quality and jeopardize the environment. As they need higher concentrations of pretreatment chemicals and have lower recovery efficiency, regions with extremely saline feed water can increase the environmental dangers associated with brine disposal.

Desalination plants raise other environmental and ecological issues in addition to the creation and disposal of brine. For instance, when the intake pumps of the desalination plant are operating, marine organisms like algae and plankton may become trapped and entrained [22]. Furthermore, the enormous amount of energy needed to run desalination facilities, which is often derived from fossil fuels, results in the production of major air pollutants such as greenhouse gases, which worsen climate change and the air quality. Desalination plants, for instance, are accountable for over a third of the greenhouse gas emissions in the United Arab Emirates (UAE). According to Alsharhan and Rizk (2020), the Intergovernmental Panel on Climate Change, 13 million m3 of drinkable water are produced using 130 million tons of oil per year, which contributes to widespread environmental contamination.

According to Areiqat and Mohamed (2005), the corals are extremely vulnerable to a rise in the Arabian Gulf's already-high seawater temperature. Other species that rely on these biotas, such as fish, are also impacted by the low water quality. The atmosphere contains both unionized (NH_3) and ionized (NH_3) ammonia. The ratio of ionized to unionized NH_3 depends on the hydrogen ion concentration (pH), and unionized NH_3 is extremely hazardous to aquatic life [23].

3. Alleviation measures

Because the applied desalination techniques, prevalent climatic conditions, types of feed water, and environmental effects of desalination plants are the same in all of the Gulf Cooperation Council (GCC) countries, there is a need for an exchange of information and expertise to solve desalination problems [24, 25]. Research on desalination has to focus on zero-brine discharge techniques such as brine processing, solar pond utilization, and the use of renewable energy sources.

4. Zero-brine discharge

The biggest environmental issue with the desalination business is rejected brine. Plants desalinating brackish groundwater have brine salinities higher than 10,000 mg/L and larger than 40,000 mg/L, respectively. According to Abdul-Wahab and Al-Weshahi (2009), brines containing at least 70,000 mg/L of total dissolved solids (TDS) are produced at 50% recovery. Chemicals for pre-, post-, and cleaning operations are also present in the brine.

The desalination method, the caliber of the feed water, the chemical additives employed during the process, and the percentage recovery all affect the physical and chemical characteristics of the rejected brine, according to Hashim and Hajjaj (2005) and Al Dousari (2009), stated that The brine produced by various desalination plants as well as the raw water and feed water is chemically analyzed. Due to the basic character of seawater, raw seawater has higher pH values, whereas feed water has the lowest pH values as a result of the addition of sulfuric acid (H_2SO_4) or hydrochloric acid (HCl) to modify the pH during pretreatment. The chemicals added during post-treatment and the mixing of the desalted water with the groundwater of various grades are to blame for the vast range of pH values of generated water [17, 26].

The fluoride ion (F) is removed from the generated water throughout the desalination process, but the rejected brine contains more F than the raw and feeds water. F might be added again to the generated water during post treatment or the blending process. TDS levels in raw water, feed water, and produced water are all high and similar, whereas produced water's TDS level is often less than 200 mg/L [27–29].

In some of the generated water, chloride (Cl^-) and sodium (Na^+) are the predominant ions, but other distillates seem to be ion-depleted water. Sulfate (SO_4^{2-}) and Ca^{2+} ions are the least prevalent, followed by bicarbonate (HCO^{3-}) and magnesium (Mg^{2+}) ions in the generated water. In all reject brines, Cl^+ and Na^+ are the two most prevalent ions. The brine contains small levels of Ca^{2+}, Mg^{2+}, and SO_4^{2-} as well.

Depending on the desalination process, feed water, and effectiveness of the desalination plants, the disposal of rejected brines from desalination plants along the Arabian Gulf coasts in Qatar, the United Arab Emirates, Bahrain, and Saudi Arabia result in the release of a variety of salts into seawater [17, 25, 30–32].

Al Asam and Rizk (2009) made the following suggestion: "Achieving zero brine discharge through brine recycling for manufacturing of various salts and chemical businesses near these plants can alleviate the negative environmental impacts of water desalination. A method for producing desalinated water and managing brine was proposed by the Dubai Electricity and Water Authority (FEWA) using a multi-phase desalination process with salinity gradient solar ponds. With this technology, the amount of created brine is zero, less energy is consumed, and solar ponds are utilized [13].

Australian studies stated that the value of brines released into the sea is believed to be six times greater than the value of the produced drinkable water [33]. Potassium chloride (KCl), magnesium chloride ($MgCl_2$), sodium chloride (NaCl), Epsom salt ($MgSO_4.7H_2O$), lithium (Li), and bromine (Br) salts are among the important salts present in the brine that has been returned to the ocean. The hypersaline brine that remains after desalination can be processed to produce salts with a marketable quality [33].

Recycled brine can be formed into chemical products like sodium hypochlorite, which is used as bleach, sodium cyanide, which is used in the gold industry, caustic soda, which is used in the aluminum industry, polyvinyl chloride, which is used in photovoltaic cells, and hydrochloric acid, which is a common acid used in all industries.

Lithium (Li), potassium (K), and bromine (Br) salts are valuable components of the brine. Li is primarily employed in the production of lithium batteries, while Br is crucial for the production of petroleum-based goods, pharmaceuticals, and a variety of fumigants. When used appropriately, brine can be a valuable resource. It can be utilized in the production of magnesium metal, Epsom salt for horticulture, lightweight flame retardant boards and panels, refractory bricks for industrial furnaces, and wastewater treatment. Brine recycling is useful in the synthesis of compounds including sodium carbonate (Na_2CO_3), sodium bicarbonate ($NaHCO_3$), magnesium chloride ($MgCl_2$), and ammonium chloride (NH_4Cl) [33].

The manufacturing of salt from the brine of saltwater desalination using reverse osmosis technology, according to Ahmed et al. (2000) and Ravizky and Nadav (2007), maybe a lucrative industry for the GCC countries. They cited several factors, including cheap desert land, very little precipitation, powerful solar radiation, quick and simple access to ports, and rather excellent accessibility to Asian countries, which are major salt consumers. Seawater's natural evaporative concentration causes dissolved salts to crystallize and precipitate throughout time in various stages. First to precipitate are calcium carbonate ($CaCO_3$) and calcium sulfate ($CaSO_4$), then sodium chloride (NaCl), magnesium (Mg), and finally potassium (K) salts. Combining reverse osmosis (RO) and Multi-Effect Solar (MES) desalination systems can increase recovery rates from 40 to 90%. The MES desalination plants preserve the brine in a closed cycle and do not release any chemicals or brine into the ocean. The brine discharged from current desalination facilities may potentially be used by the MES plants to produce drinking water [34–36].

5. Technology for solar ponds

Al Asam and Rizk (2009) stated that, "Flat solar radiation land and water are widely available, making solar ponds as a source of renewable energy desirable. Solar ponds require year-round solar exposure, significant quantities of brine, a sufficient supply of "fresher" water, inexpensive flat terrain with low permeability, and steady electricity demand to be effective electricity generators. The degree of brine purity, the thickness of the layers within the solar pond, the upkeep of the vertical salt gradient, the area of the pond, and the depth of the groundwater all have an impact on the thermal efficiency of these systems. Normally, when water is heated, it rises to the surface, but solar ponds prevent this from happening because a lot of salt is dissolved in the hot bottom layer of the pond, making the water too dense to rise. The lower layer of the solar pond is hot (70–100°C) and has a very high salinity, whereas the upper layer is cool and has a low salt content [13, 37].

Water in the gradient zone cannot rise because the upper water is lighter and has a lower salinity, noted Safi and Korchani (1999). The water underneath has a higher salinity and is heavier, thus the water on top cannot travel downward. To allow sunlight to be trapped in the warm bottom layer, which may then be used to generate heat energy, the gradient zone can function as a transparent insulator. Solar ponds with a salinity gradient capture and store solar energy during the day and release it for desalination at night. In the pond's lowest level, solar energy is received and transformed into thermal energy. The thermal energy can subsequently be applied to the production of electricity, desalination, and space heating [37].

Because the solar pond serves as both a collection and storage mechanism, it has an advantage over other solar energy collection techniques. As a result, the pond can continue to produce electricity even when the sun is not out. The solar-pond technique needs a vast, inexpensive land area, a lot of solar energy, and a company that can use hot water effectively to offset the cost to be practical. The salinity-gradient solar-pond method for saltwater desalination along the beaches appears to be highly feasible given these circumstances.

Direct - desalination, where thermal desalination takes place in the same device, and indirect solar desalination, where the plant is divided into the solar collector and the desalination system, are the two types of solar water desalination systems [11].

The solar pond collector has the following advantages [38–40]:

- Solar ponds can have a simple design, a very large heat-collection area and low cost.
- The heat storage is massive and enables energy extraction day and night.
- The major production potential is during peak electrical-power demand in the middle of the summer.
- Generation of the heat required for domestic, industrial and agricultural applications, such as desalination, heating and generation of electricity.
- The technology and scientific principles are well understood and well documented in scientific papers.
- Production of the energy needed for the production of salts for commercial uses and reduction of the emission of greenhouse gases.

6. Renewable energy usage

Solar-powered desalination facilities in the GCC nations were the subject of several studies, including Hanafi (1991), Trieb (2007), and DLR (2007). However, except for pilot plants in a few nations, which are primarily for research purposes, there are no significant attempts to build large-scale solar desalination plants [40–42]. The first solar desalination plant in Umm Al Nar, Abu Dhabi, was created to determine if sun desalination was practical in dry regions [43, 44]. The plant uses groundwater that is saltier than seawater, has a daily capacity of 15,000 gallons (GPD), and is powered by photovoltaic panels. On Sir Bani Yas Island in the Abu Dhabi Emirate, the German company SYNLIFT Systems began operating a wind-powered saltwater desalination facility in 2003 [45]. By taking into account the type of energy that is primarily needed to run the operation, another helpful classification can be realized.

This factor is crucial for the supply of the desalination process from specific renewable energy sources (**Figure 2**) [46, 47]. Four types of energy are specifically taken into account here:

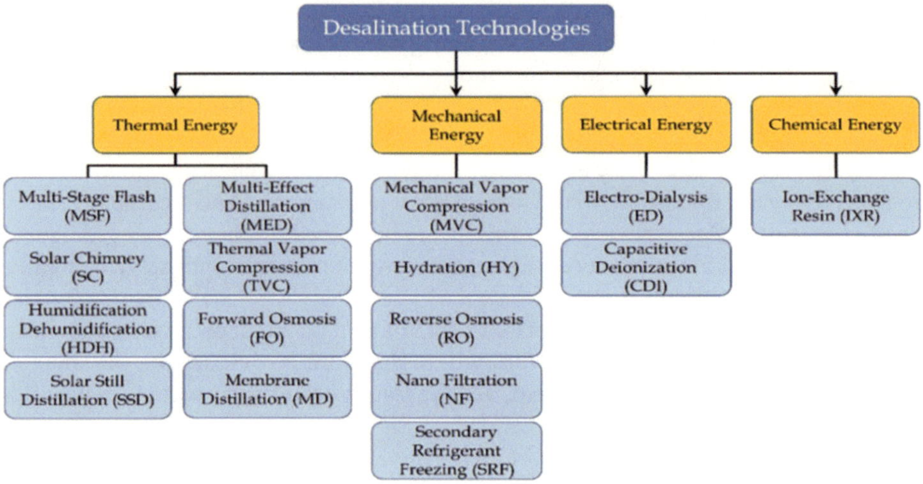

Figure 2.
Shows desalination methods are categorized according to their primary energy input.

- Thermal energy.
- Mechanical energy.
- Electrical energy.
- Chemical energy.

Therefore, it is useful to make a distinction between energy sources that can be used to produce thermal energy and electricity to provide the desalination process with renewable energy [43]. According to the typical energy output that may be produced, renewable energy sources can be categorized in the following ways to achieve this goal (**Figure 3**) [48]:

- Thermal and electrical energy producers, such as solar, geothermal, and biomass.
- Electricity producers, such as wind, hydro, tidal, and wave. The local energy resource's characteristics are typically taken into consideration while choosing the energy output.

It is crucial to solve the issue of brine generation in desalination plants. Even while the brine is frequently sufficiently diluted before being dumped back into the ocean, it is still feasible that even a small variation from the typical salinity levels will have an impact on marine life and environments. It was originally believed that the ocean was too big for anthropogenic activity to have a substantial impact, however problems like ocean acidification show that this is far from the case and even minor cumulative inputs of toxins can have large-scale effects. The creation and operation of large-scale projects like desalination plants, which can have so many negative effects, requires prudence [11, 49–51].

Finally, desalination technology offers enormous potential for supplying water to a burgeoning global population. The need for freshwater will be met, water security will be improved, groundwater mining will be reduced, and issues with

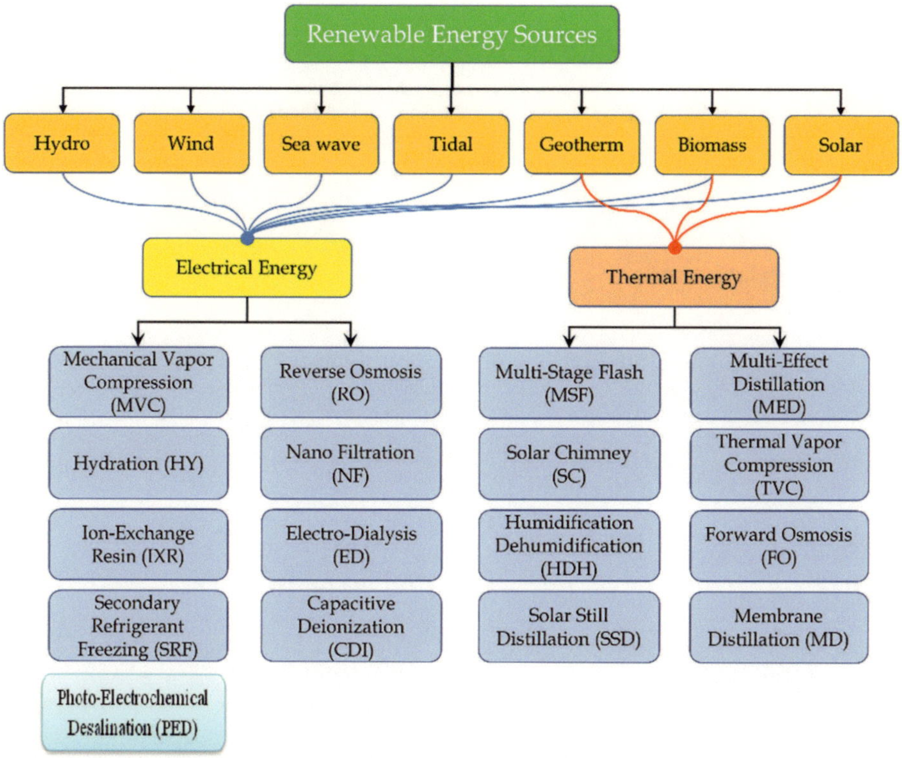

Figure 3.
Shows desalination technologies and renewable energy sources that might be combined.

public health brought on by consuming tainted surface water will be lessened. Even so, it might help ease friction over water allocation rights between and within nations. The technology must therefore be developed further, but we must also work to reduce the specific environmental and health effects it has. Desalination will become a more affordable and sustainable option for supplying water to the growing global population as a result of improved brine discharge control and desalination plant efficiency.

Author details

Adel Hussein Abouzied
Central Laboratories, The Holding Company for Water and Waste-Water Treatment, Giza, Egypt

*Address all correspondence to: adelabouzied@gmail.com

References

[1] MOEW (Ministry of Environment and Water). Water Conservation Strategy. UAE (United Arab Emirates): Ministry of Environment and Water; 2010. p. 212

[2] Gonzalez J, Cabrera P, Carta JA. Wind energy powered desalination systems. In: Desalination: Water from Water. 2nd ed. Hoboken, NJ: Wiley; 2019. pp. 567-646

[3] MOEW. State of the Environment Report—United Arab Emirates. UAE (United Arab Emirates): Ministry of Environment and Water; 2015. p. 36

[4] Shahzad MW, Burhan M, Ng KC. Pushing desalination recovery to the maximum limit: Membrane and thermal processes integration. Desalination. 2017;**416**:54-64

[5] Ibrahim AGM, Rashad AM, Dincer I. Exergoeconomic analysis for cost optimization of a solar distillation system. Solar Energy. 2017;**151**:22-32

[6] Rosales-Asensio E, Garcia-Moya FJ, Borge-Diez D, Colmenar-Santos A. Sea water desalination in micro-grids. In: Green Energy and Technology. Ali Sadeghi Digital Library. Springer; 2022. pp. 112-130

[7] Tzen E, Rossis K, Gonzalez J, Cabrera P, Penate B, Subiela V. Wind technology design and reverse osmosis systems for off-grid and grid-connected application. In: Mahmoudi desalination. 1st ed. London: CRC Press; 2017

[8] Carta JA, Gonzalez J, Cabrera P, Subiela VJ. Preliminary experimental analysis of a small-scale prototype SWRO desalination plant, designed for continuous adjustment of its energy consumption to the wilely varying power generated by a stand-alone wind turbine. Applied Energy. 2015;**137**:222-239

[9] Qasim M, Badrelzaman M, Darwish NN, Darwish NA, Hilal N. Reverse osmosis desalination: A state-of-the-art review. Desalination. 2019;**459**:59-104

[10] Mona M, Amin A, Ghada A. Water Treatment and Desalination. London, UK, London, UK: IntechOpen; 2020. DOI: 10.5772/intechopen.91471

[11] Curto D, Franzitta V, Guercio A. A review of the water desalination technologies. Applied Sciences. 2021;**11**:670. DOI: 10.3390/app11020670

[12] Dickie P. Desalination: Option or Destruction for a Thirsty World? Report Prepared for WWF's Global Freshwater Program. Switzerland: WWF International; 2007. p. 53

[13] Al Asam MS, Rizk ZE. Desalination and Water Environment in the United Arab Emirates: Impacts and Solutions. Dubai, UAE: International Desalination World Congress; 2009. p. 41

[14] Younos T. Environmental issues of desalination. Journal of Contemporary Water Research & Education. 2005;**132**:11-18

[15] Al Barwani HH, Purnama A. Evaluating the effect of producing desalinated seawater on hypersaline Arabian gulf. European Journal of Scientific Research. 2008;**22**(2):279-285

[16] Höpner T, Lattemann S. Chemical impacts from seawater desalination plants – A case study of the northern Red Sea. Desalination. 2002;**152**:133-140

[17] Al Dousari AE. Desalination leading to salinity variations in

Kuwait marine waters. American Journal of Environmental Sciences. 2009;**5**(3):451-454

[18] Frank H, Rahav E, Bar-Zeev E. Short-term effects of SWRO desalination brine on benthic heterotrophic microbial communities. Desalination. 2017;**417**:52-59

[19] Kämpf J, Clarke B. How robust is the environmental impact assessment process in South Australia? Behind the scenes of the Adelaide seawater desalination project. Marine Policy. 2013;**38**:500-506

[20] DeNicola E, Aburizaiza OS, Siddique A, Khwaja H, Carpenter DO. Climate change and water scarcity: The case of Saudi Arabia. Annals of Global Health. 2015;**81**(3):342-353

[21] Russell D. Dissolved air flotation and techniques. In: Practical Wastewater Treatment. Wiley Online Library; 2019. pp. 409-417. DOI: 10.1002/9781119527114.ch19

[22] Blanco-Marigota AM, Lozano-Medina A, Marcos JD. The exergetic afficiency as a performance evaluation tool in reverse osmosis desalination plants in operation. Desalination. 2017;**413**:19-28

[23] Areiqat A, Mohamed KA. Optimization of the negative impact of power and desalination plants on the ecosystem. Desalination. 2005;**185**:95-103

[24] ESCWA. Development of Frameworks to Implement National Strategies of Integrated Water Resources Management in the ESCWA Countries. New York: United Nations; 2005. p. 94 (in Arabic)

[25] FEWA (Federal Electricity and Water Authority). FEWA Annual Statistical Report. Ajman, UAE: Etihad Water and Electricity Open Data Policy; 2009

[26] Hashim A, Hajjaj M. Impact of desalination plants fluid effluents on the integrity of seawater: With the Arabian gulf in perspective. Desalination. 2005;**182**:373-393

[27] WHO (World Health Organization). WHO Guidelines for Drinking Water Quality, Recommendations. Vol. 1. Geneva, Switzerland: WHO; 1984. p. 130

[28] Cotruvo JA, Abouzaid H. New World Health Organization Guidance for Desalination. Health and Environmental Impacts. Cairo: WHO Eastern Mediterranean Regional Office; 2007. p. 173

[29] Rizk ZS. Inorganic chemicals in drinking water of the United Arab Emirates. Environmental Geochemistry and Health. 2009;**31**:27-45

[30] Al Mutaz IS, Al SB. Operation characteristics of Maufouha reserve osmosis plants. In: The IDA World Congress on Desalination and Water Reuse, 6-9 October 1997. Madrid, Spain: The IDA World Congress on Desalination and Water; 1997. p. 25

[31] Hassan MH. Major distribution in Khobar aquifer, Eastern Saudi Arabia. Arabian Journal for Science and Engineering. 1998;**23**(1C):67-88

[32] Al Noaimi MA. Evaluation of available water resources and their uses in the state of Bahrain: Bahrain Center for Studies and Research. Series of Studies and Scientific Research Papers. 1999;**24**:131 (in Arabic)

[33] Ibrahim AA, Jibril BY. Production of chemicals from desalination blow down. Alexandria Engineering Journal. 2005;**44**(5):817-820

[34] Ahmed M, Arakel A, Hoey D, Coleman M. Integrated Power, Water and

Salt Generation—A Discussion Paper. Oman: Department of Soil and Water, Sultan Qaboos University; 2000. p. 11

[35] Ravizky A, Nadav N. Salt production by evaporation of SWRO brine in Eilat: A success story. Desalination. 2007;**205**:374-379

[36] Pereira MC, Meddes JF, Horta P. Advanced solar dryer for salt recovery from brine effluent of desalination MED plant. In: Solar Energy a Sustainable Future. Sweden: ISES Solar World Congress; 2003. pp. 16-19

[37] Safi MJ, Korchani A. Cogeneration applied to water desalination: Simulation of different technologies. Desalination. 1999;**125**:223-229

[38] Lu H, Walton JC, Swift AHP. Desalination coupled with salinity-gradient solar pond. Desalination. 2001;**136**:13-23

[39] Lu H, Walton C, Hein H, Water Treatment Engineering and Research Group. Thermal Desalination Using MEMS and Salinity-Gradient Solar Pond Technology. USA: US Department of the Interior, Bureau of Reclamation; 2002. p. 91

[40] Hanafi A. Design and performance of solar MSF desalination system. Desalination. 1991;**82**(1-3):165-174

[41] Trieb F. Concentrating Solar Power for Seawater Desalination. Damascus, Syria: MENAREC 4; 2007. p. 12

[42] DLR (German Aerospace Center). Concentrating Solar Power for Seawater Desalination. Germany: Federal Ministry for the Environment, Nature Conservation and Nuclear Safety; 2007. p. 12

[43] Wei QJ, McGovern RK, Lienhard VJH. Saving energy with an optimized two-stage reverse osmosis system. Environ. Sci.: Water Resour. Technology. 2017;**3**:659-670

[44] Al-Nashar AM. Effect of dust deposition on the performance of a solar desalination plant operating in an arid desert area. Solar Energy. 2003;**75**(5):421-431

[45] Käufler J. Experience in Morocco and Abu Dhabi (UAE). Berlin, Germany: SYNLIF Systems GmbH; 2007. p. 31

[46] Ma Q, Lu H. Wind energy technologies integrated with desalination systems: Review and state-of-the-art. Desalination. 2011;**277**:274-280

[47] Alkaisi A, Mossad R, Sharifian-Barforoush A. A review of the water desalination systems integrated with renewable energy. Energy Procedia. 2017;**110**:268-274

[48] Goosen M, Mahmoudi H, Ghaffour N. Water desalination using geothermal energy. Energies. 2010;**3**:1423-1442

[49] ICBA (International Center for Biosaline Agriculture). Developing Federal Environmental Guidelines and Standards to Monitor and Manage the Discharges from Desalination Plants in the United Arab Emirates. UAE: International Center for Biosaline Agriculture, UAE Ministry of Environment and Water; 2012. p. 119

[50] Abdul-Wahab SA, Al-Weshahi MA. Brine management: Substituting chlorine with on-site produced sodium hypochlorite for environmentally improved desalination processes. Water Resources Management. 2009;**23**:2437-2454

[51] Alsharhan AS, Rizk ZE. In: Singh VP, editor. Water Resources and Integrated Management of the United Arab Emirates. World Water Resources. Vol. 3. USA: Department of Biological and Agricultural Engineering & Zachry Department of Civil Engineering, Texas A&M University; 2020. DOI: 10.1007/978-3-030-31684-6

Chapter 5

Perspective Chapter: Hydrogel Draw Agent Desalination Systems – Outlook

Alexander Fayer

Abstract

The chapter intends to discuss an application of hydrogel material as draw agent for a forward osmosis desalination system. This refers to systems that allow a continuous process of extraction of desalinated water with low energy costs and minimal environmental pollution. One of the most prominent properties of hydrogel materials is their ability to spontaneously absorb large quantities of water from saline solution separated by a semipermeable membrane. This process is energetically favorable due to the difference in the chemical potentials of water in the solution and hydrogel. Thermodynamic equilibrium between hydrogel and external saline solution corresponds to the strictly defined amount of water retained by the hydrogel in the given conditions. The excess pressure of water in hydrogel relative to the pressure of the pure external in this state is defined as the osmotic pressure difference. In contrast to the absorption of water molecules by hydrogel, their extraction is usually a process that requires large energy consumption and disruption of the continuity of the desalination cycle. However, known several opportunities to overcome this bottleneck and they are discussed in detail.

Keywords: hydrogel, swelling, water extraction, forward osmosis, draw agent, desalination, wicking, solar powered heating

1. Introduction

Osmosis-based water desalination is an effective technique to produce high-quality water and the production rates are easily adjustable. Although two types of the osmosis-based desalination, namely forward osmosis (FO) and reverse osmosis (RO), are known currently, only the RO continuous process has found wide industrial implementation and its market share in water desalination is rapidly increasing. However, the RO based water desalination is generally recognized as energy-intensive (2.2–3.5 kWh/m^3) process [1]. The high energy usage during RO desalination causes environmental concerns such as air pollution and heating associated with water cooling using energy production from fossil fuels. Another drawback of RO-based desalination is the production of high-salinity brine, which contains plenty of substances and chemicals that are harmful to the environment and ecosystem. Several studies have suggested solutions to reduce the high energy consumption and brine impacts. However, the industry community is developing the opinion that RO-based desalination has reached the theoretical and practical limit. The energy limitation and environmental damage must be overcome through different technical solutions [2]. One of desalination techniques considered as the possible alternative is forward osmosis (FO) process. In contrast to RO, in which work is done to push water molecules through the membrane against a pressure drop, FO is an energetically favorable process due to the difference in the chemical potentials of water in the solution and in draw agent. Water molecules accommodated by draw agents have different liquid water properties depending on the interaction within the agent material. Diffusion of water through the membrane is affected by the level of the agent hydration. Thermodynamic equilibrium between draw agent and external aqueous solution corresponds to the strictly defined amount of water retained by the draw agent in given conditions. A successful FO desalination process is critically reliant on the availability of a draw agent that offers both high osmotic pressure and a facile regeneration mechanism. It is generally accepted that FO process inevitably requires post-process for all types of draw agents to obtain the final water production, which creates significant obstacles to its use as a stand-alone low-energy desalination process and commercialization [3].

2. Peculiarities of hydrogel-based FO desalination process

Hydrogels are crosslinked three-dimensional hydrophilic polymer networks that can absorb a huge amount of water and not be dissolved in it. Some hydrogels called smart or stimuli-responsive polymer hydrogels can undergo a reversible volume change or solution-gel phase transition in response to external environmental stimuli, including temperature, light, pressure, solvent composition, and pH. These intrinsic properties led to search among plenty of hydrogels for such that have lowest regeneration energy consumption and correspond to such requirements as high osmotic pressure, be nontoxic, exhibit low reverse flux and acceptable cost [4]. Various hydrogels were studied as draw agents in the FO process over the past few years with varying degrees of success. The results of these efforts have been summarized by Wang group and presented in review [5]. However, as noted in the review, there are some issues that still need to be addressed in hydrogels application as draw agents, including low water flux, high external concentration polarization, and especially the in-continuous operation. Overcoming of the last drawback is of primary complexity. Let us briefly consider basics of functioning of the FO desalination system permanently extracting water from a hydrogel draw agent. In order to facilitate the description, but without limiting the generality, we will consider a notional one-dimensional water desalination processes that can take place in a vessel divided into three parts. Let the first third of the vessel be filled with running saline water and separated from the middle part by a semipermeable membrane for FO desalination. The middle part is crowded with granular unsaturated hydrogel material and separated by grid from the third part of the vessel, which is intended for collection of the freshwater. Saline water pressure P_f at upper vessel's part does not exceed pressure values in ordinary drinking water supply systems that is of about 0.5 MPa. · Only water molecules penetrate through the membrane and cause swelling of the constrained hydrogel material. The swelling leads to an increase of hydraulic pressure and probably efflux of part of water through the wick starting from some pressure value. Hydrogels consist of two phases, the polymer network, which is constant in quantity, and the aqueous phase, which is variable. The system under consideration would not reach the state of thermodynamic equilibrium between these phases if a part of the desalinated water flux entering the hydrogel is diverted through the grid to the third part of the vessel under the influence of some factors. In what follows we will consider only two such factors, specially selected wicks and solar radiation. An implementation of the described above scheme imposes number of specific requirements for the properties of used materials. These requirements are discussed below.

2.1 Hydration of hydrogels

One of the most prominent properties of hydrogels is their ability to absorb large quantity of water and to swell as a consequence of this process. Water molecules accommodated by hydrogels have different properties depending on the position and interactions within the hydrophilic network. A model, presented in 1973 by John Andrade [6], defines three types of water in hydrogels—nonfreezing or bound water, free or bulk water, and freezing interfacial or intermediate water. Molecules of free water are not affected by the polymer and freeze/melt similarly to pure water; molecules of bound and to some degree intermediate water are immobilized by binding to the polymer chains through hydrogen bonds. Other properties of the bound and the intermediate water (relaxation time, polarization, etc.) also differ from properties of free water.

Diffusion in hydrogels is affected by the level of hydration. Experiments with tracer molecules dissolved in water elucidate that at a high levels of hydration, the process occurs primarily as the free water diffusion; however, at low hydration, it takes place as diffusion in the bound water [7].

Note that the thermodynamic equilibrium between hydrogel and external aqueous solution corresponds to the strictly defined amount of all three types of water retained by the hydrogel in the given conditions. The excess pressure of water in hydrogel relative to the pressure of the pure external in this state is defined as the osmotic pressure difference and can be expressed through the difference of the corresponding chemical potentials [8, 9] for ideal elastomeric gels. The hydrogel's osmotic pressure, which can be called the "driving force" of the hydration process, is a complex parameter. The total osmotic pressure in a swelling hydrogel Π can be divided into three separate parts using the Flory–Rehner theory. This theory [10] states the perfect separability of the total free energy, (ΔF), into an elastic, mixing, and ionic contributions, each with an associated osmotic pressure (Π_{elastic}, Π_{mixing}, and Π_{ionic}).

$$\Pi = \Pi_{\text{ionic}} + \Pi_{\text{mixing}} + \Pi_{\text{elastic}} \tag{1}$$

The mixing and ionic contributions are commonly seen as the cause of gel swelling while the elastic portion restricts the large expansion of the material.

The mixing osmotic pressure refers to the attraction of solvent molecules in the external solution to the hydrophilic polymer chains and can be expressed [11] by the Flory–Huggins

$$\Pi_{\text{mixing}} = -\left[\ln(1-\varphi) + \varphi + \mathit{T}\varphi^2\right] \tag{2}$$

where φ is the molar volume of the solvent, R is the universal gas constant, T is the temperature (Kelvin), V is the current solid volume fraction, and T is Flory–Huggins parameter derived from the solid-fluid interaction. This parameter is material and environment dependent and defines deswelling properties of hydrogels.

For χ > 0.5 the hydrogel solution is unstable for small fluctuations [12] and gives off water relatively easily.

The equilibrium pressure for real hydrogels is different from the osmotic pressure and refers here as osmotic swelling pressure. Its value as well as equilibrium swelling ratio and swelling kinetics differs for free-standing and confined hydrogels.

A steady state water flow if such established along the porous medium like confined hydrogel can be described by following equation

$$Q = \frac{SoK(\Delta \Pi - \Delta P)}{d\mu} \tag{3}$$

where Q is the rate of water flow, K is the hydraulic conductivity of the hydrogel block, d is the hydrogel layer thickness, S_0 is the hydrogel cross-section, μ is the solvent viscosity, and ΔP and $\Delta \Pi$ are the hydraulic and osmotic pressures difference between the input and the output edges of the hydrogel block, respectively.

Note that only movement of unbounded water contributes to flow, while liquid molecules held by absorptive forces are essentially immobile [13].

2.2 Forward osmosis membrane assembly—hydrogel interface

A flow of water through a semipermeable membrane can be described by Darcy's law in its complete form [14]:

$$Qm = \frac{SDwCm}{\lambda}\left\{1 - exp\left[\frac{Vw}{RT}(\Delta\pi - \Delta Pm)\right]\right\} \tag{4}$$

where Q_m is the rate of water flow, S is the effective cross-section of membrane, D_w is the average water diffusion coefficient in the membrane, C_m is the equilibrium concentration of water in the membrane, λ is the membrane thickness, V_w is the partial molar volume of water, R is universal gas constant, T is the temperature (Kelvin), ΔP is the pressure difference across the membrane assembly, that is, $\Delta P_m = P_p - P_f$, where P_p is hydraulic pressure at the cross-section of hydrogel located at the distance of free path of water molecules from the membrane assembly and P_f is the feed pressure, and $\Delta\pi$ is the "the driving force" of water flow that is difference of osmotic pressures Π_p and Π_f at the above-mentioned cross section and the feed solution, respectively, while Πf directly proportional to the molality a of the solution M:

$$\Pi_f = MRT \tag{5}$$

Carr [15] defined the characteristic timescale for a diffusion process τ as the maximum value of the mean action time across the layer

$$\tau = 0.5\frac{l^2}{D} \tag{6}$$

where l is the layer's thickness and D is the coefficient of diffusion.

The characteristic timescale of water diffusion in hydrogel block exceeds by many orders of magnitude, the same parameter for the membrane assembly. As a result, a gel layer with a thickness approximately equal to the pore size in a rigid membrane's base and in close proximity to it approaches the state of local equilibrium with the feeding solution however does not reach it [16]. In this, the water flow through the membrane decreases significantly compared to its value in the absence of a hydrogel (effect of the concentration polarization). Because osmotic flow and hydraulic flow require the same pressure drop along the membrane pore to generate equal flow [17], this effect can be expressed by system of Eq. (7)

$$\begin{cases} \mathbf{\Pi_p = \Pi_f - \Delta\Pi_{eff}} \\ \mathbf{P_p = P_f} \end{cases} \tag{7}$$

A magnitude of $\Delta\Pi_{eff}$ depends on the properties of both the membrane assembly and the hydrogel and lies in the range of 1200–2200 KPa [18].

Considering the effect of the concentration polarization relations given in Eq. 5 for steady-state water flow along the hydrogel block can be rewritten as:

$$Q = SoK \tag{8}$$

Since the resistance of the hydrogel block is at least four orders of magnitude greater than the analogous parameter of the membrane assembly (an asymmetric membrane normally consists of a dense layer of 0.1–1 μm thick and supported by a highly porous, 100–200 μm thick support layer [19]), the same equation can be used to describe the water flow through complete membrane-hydrogel block subsystem.

Based on Eq. (8), water flow is decreasing function of P_l and reaches zero at its ceiling value (or upper limit) P_{lmax}

$$\mathrm{P_{lmax} = \Pi_l - \Pi_f + \Delta\Pi_{eff} + P_f} \tag{9}$$

3. FO continuous desalination by wicking of pH-sensitive hydrogel agent

Freshwater recovery is a major embarrassment in direct osmosis desalination technology in general and hydrogel-based FO desalination in particular.

One of the possible ways to provide an energy-efficient process with a continuous duty cycle to overcome the above bottleneck is proposed and experimentally tested in the article [20]. The idea was to provide local stimuli impact on grains of pH-sensitive hydrogel. **Figure 1** illustrates typical swelling dependency of superabsorbent hydrogel on pH value. As follows from the graph the water content in the hydrogel reaches a maximum at a certain pH value, and any change in this value may be accompanied by spontaneous emission of water.

The local release of water from hydrogels under such factors as laser pulse and mechanical puncture was observed in works in ref. [22, 23]. However, these methods cannot provide collection of the released water.

The authors of ref. [20] proposed to use wicks with surface pH different from the pH of hydrogel medium as stimulus for local release of water from hydrogel granules into the intergranular space. This construction proved to be capable of passive extraction of water from swelling hydrogel draw agent in three-stage process:

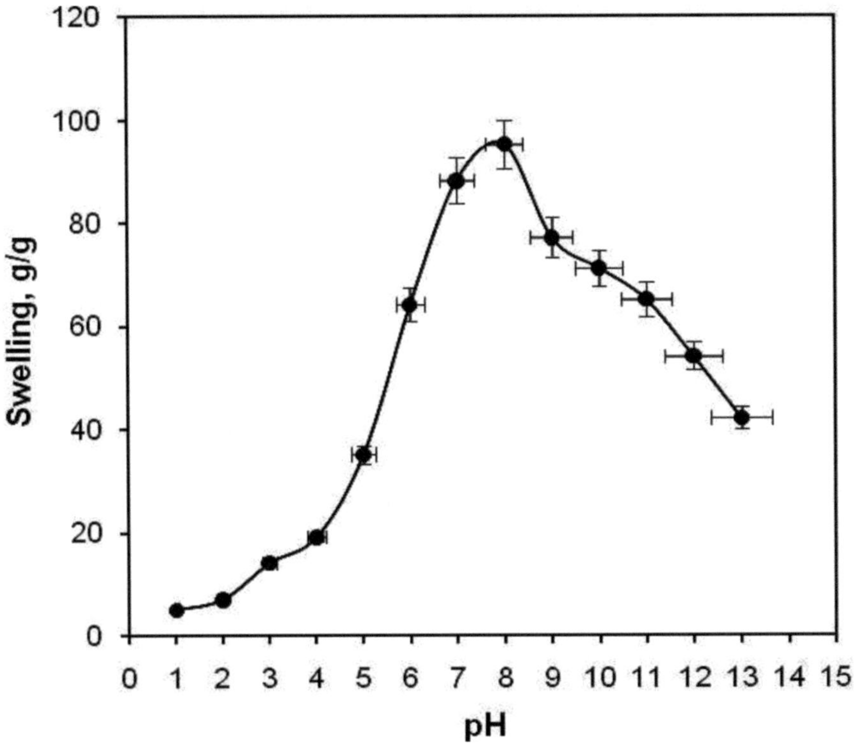

Figure 1.
pH-dependent swelling of the superabsorbent hydrogel (according to [21]).

- efflux of water from hydrogel grains into intergranular space in the interface with embedded wicks,
- soaking up intergranular water with the wicks, and
- dripping or evaporation of water from the wicks.

Wicking is a spontaneous movement of liquids into porous media under the action of the capillary suction pressure. Value of the suction forces is governed by the properties of the liquid, liquid-medium surface interactions, and geometric configurations of the pore structure in the medium.

At the conditions of steady-state, all liquid entering the wick per unit of time will leave it in the same period by the drop flow (evaporation-protected wicks).

Recently invented types of wicks can drain freshwater with corresponding suction pressure P_{sw} as high as hundreds of kilopascals.

The lowest hydraulic pressure P_t ("threshold pressure") starting from which the water enters the wick is determined from the balance of promoting and hindering forces acting on an element of free water at the hydrogel-wick interface. Generally, this pressure is unattainable if you try to extract water directly from hydrogel grains since water retaining component of the osmotic pressure ("suction pressure of hydrogel"), which far exceeds P_{sw} value. However, intergranular water initiated by pH-difference is easily extractable with wicks as demonstrated in experimental results of Ref. [20] shortly presented below.

A verification of water extraction feasibility from swelled hydrogel preceded by complete removal of liquid from the intergranular space of the washed hydrogel by multistep procedure using equilibration solutions of different salinity separated from hydrogel by FO membrane.

The effect of the wick's surface pH-initiated water release was investigated by simultaneous immersing of one end of each of the two test wicks into the bicker filled with potassium super absorbent polymer hydrogel while the other end of the wicks hung loosely down. The phenomenon of water extraction by wicks has been detected experimentally by measurement of liquid front propagation rate at various pH values of the wick and hydrogel media as well as the salinity of equilibration solution. The measurement of average rate of the liquid front advancement along the wicks has been performed by optical image analysis method [24] (the wicks were protected from water evaporation).

Figure 2 reflects an exponential drop in the rate of the waterfront advancement occurring with an increase in the concentration of the equilibrating solution and a corresponding decrease in the amount of water in the hydrogel characterized by pH value equal to 6.5.

Two wicks 1 and 2 differing in the values of their surface pH (7.8 and 7.2 respectively) were used in this experiment. As can be seen from the same graph, the rate of water extraction by the first wick W1 is higher than by the second one (W1 – W2 > 0). There are two possible reasons for this phenomenon: inequality of the suction forces of two wicks and inequality of local hydrogel shrinking because of the wick's pH difference. In the first case the sign of difference (W1 – W2) is independent but in the second case must be dependent on the hydrogel pH value change in a certain range. The results of the experiment with a change in the pH value of the hydrogel are presented in **Figure 3**.

About the same ratio between the extraction rates is maintained for hydrogel with pH equal to 6.5 and 6.9. However, the test provided with hydrogel whose pH was 7.6 resulted in a change in the sign of the difference (W1 – W2 < 0).

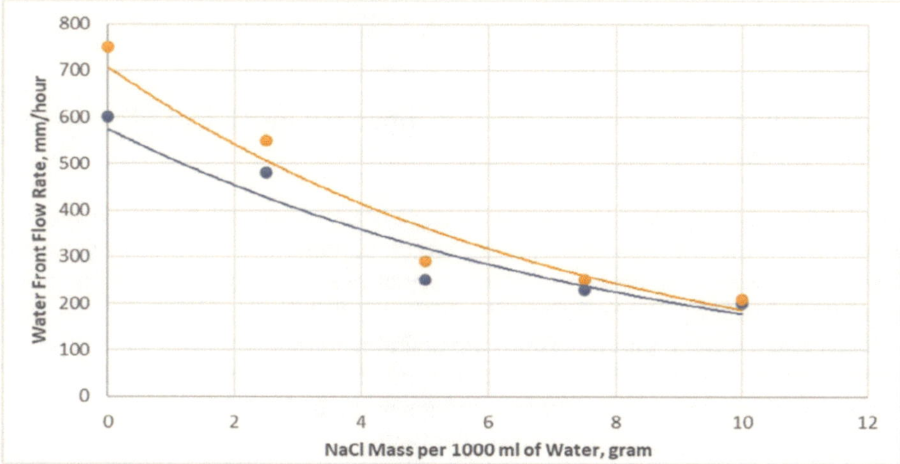

Figure 2.
Waterfront rate as function NaCl content in the equilibration solution for the wicks with surface pH values 7.8 (red line) and 7.2 (blue line).

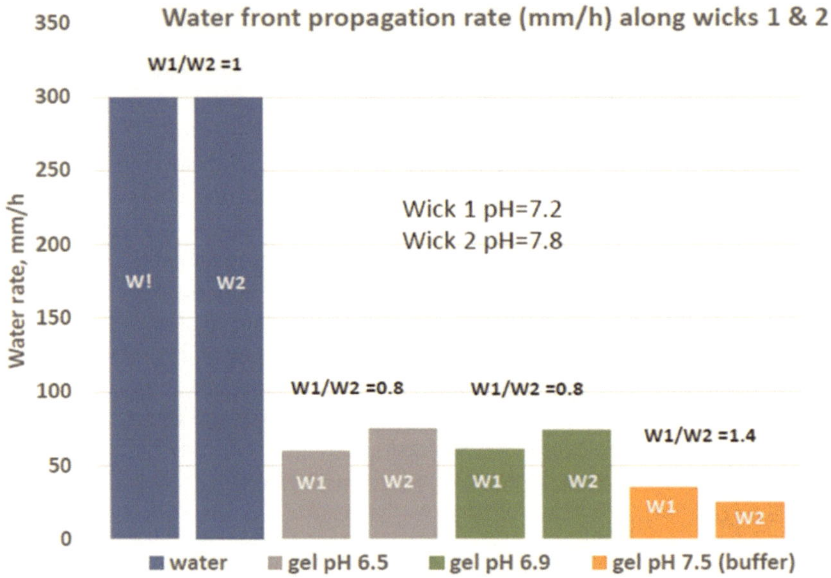

Figure 3.
Rate of waterfront propagation along wicks 1 and 2 inserted in hydrogels equilibrated with distilled water and having different pH values.

The effects described above have been confirmed by exploitation of a prototype of continuously operating FO desalination system providingspontaneous flow of fresh-water outflow from the container with saline water and consistently passing through the semi-permeable membrane, the hydrogel block, and system of the specified wicks. At conditions where the salinity of the source water is in the range of 0–10 g/l, the potassium polyacrylate hydrogel acts as a "water pump" and a "water bridge" simultaneously. This phenomenon can be used for desalination of underground water for needs of irrigation whereas desalination of sea water requires application of pH-sensitive hydrogels with higher osmotic pressure.

4. FO continuous desalination by solar-powered heating of pH-sensitive hydrogel agent

Desalination of seawater by solar energy is a hot topic of hydrogel draw agent concept. In the routine solar dewatering process, polymer hydrogels deswell under solar-induced heating resulting in the recovery of pure water and the recycling of composite polymer hydrogels for another FO process. The core element of the composite draw agent is the photothermal conversion materials like black carbon particles or more sophisticated like carbon nanotubes and aluminum-based plasmonic absorbers for example. The hydrophilic groups in the hydrophilic polymer network are beneficial in reducing the evaporation enthalpy of water molecules, accelerating water transport and improving the solar energy conversion efficiency [25, 26]. Wang et al. for the first time proposed the intermittent stimuli impact on hydrogel draw agent for quasi-continuous production of freshwater by using heating by solar energy-cooling cycles. Actually, they proposed to replace uniform bulk hydrogel body by a

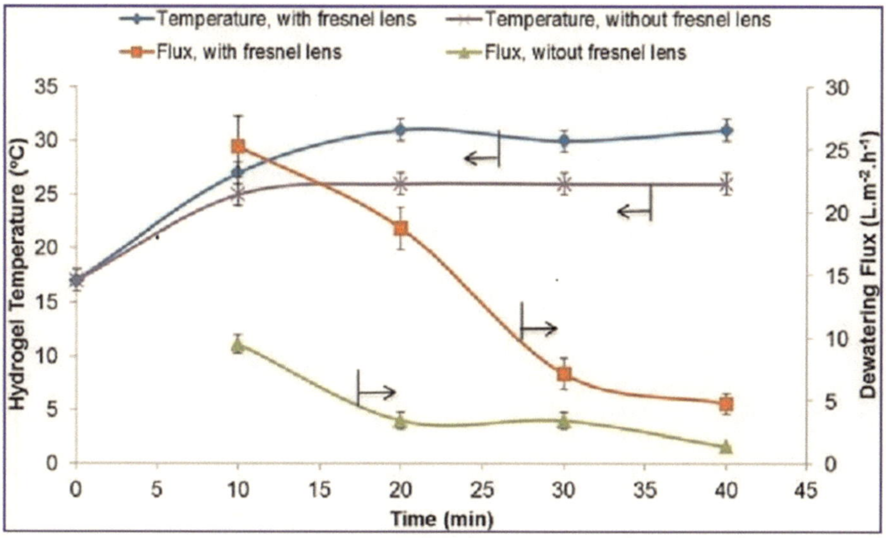

Figure 4.
Schematic illustration of bifunctional polymer hydrogel layers process [27]. Dewatering flux of thermoresponsive hydrogel as draw agent in the bilayer arrangement (solar intensity = 0.5 kW/m² (W_{input} = 2 kW/m²), Q = 15, 0.2 g dewatering layer, 0.01 g absorptive layer).

bilayer structure with no need to remove the draw agent from the membrane module [5, 27]. The feasibility of bilayer polymer hydrogels as draw agents in FO process has been investigated. The dual-functionality hydrogels consist of a water-absorptive layer to provide osmotic pressure, and a dewatering layer to allow the ready release of the water absorbed during the FO drawing process at lower critical solution temperature (LCST) (**Figure 4**).

Few years later the idea of Wang group was amplified by Chen et al. [28]. They developed laminated temperature-responsive hydrogel based on poly(*N*-isopropylacrylamide-co-sodium acrylate) (P(NIPAAm-co-SA)) with variable content of SA. The concentration of sodium acrylate decreased away from the FO membrane in the drawing layer, and the releasing layer was pure PNIPAm. Adding intermediate layers or employing the intermittent dewatering strategy increased the dewatering ratio for the multilayer hydrogel because water molecules could be transported from the drawing layer to the releasing layer more easily. The design illustrated in **Figure 5** can effectively decrease reverse osmotic pressure, resulting in an increase in water flux.

The multi-layer hydrogel released ~60% of the absorbed water at the fully swollen state in 60 min due to the LCST phase transition of the P-NIPAAm releasing layer while the uniform hydrogels only released ~35% of the absorbed water purely by evaporation. The FO flux of multilayer hydrogel is still low compared to the inorganic hydrogel, which is a key obstacle for most organic draw agents. However, the multilayer temperature-responsive hydrogel had a low energy consumption compared to other regeneration methods for nonresponsive draw agents due to the LCST phenomenon, which was rid of the latent heat penalty. The concept of multilayer design showed promising application by replacing the releasing layer with highly ionic polyelectrolyte [28].

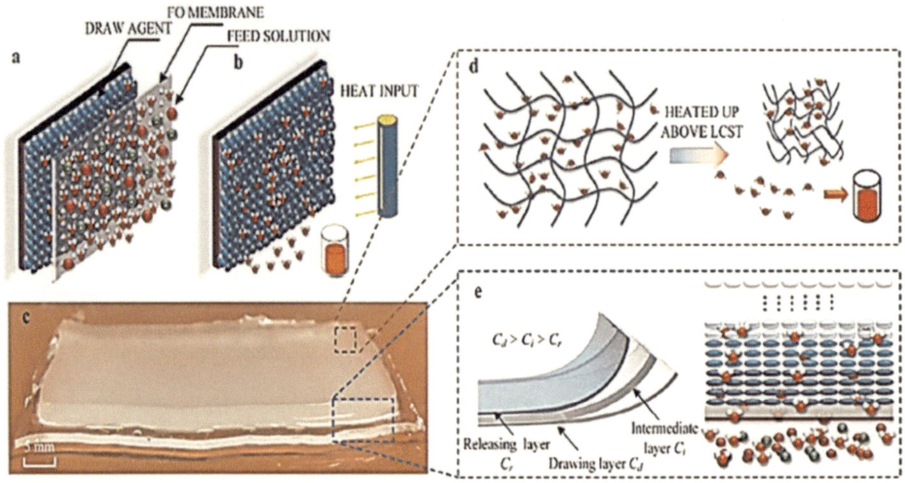

Figure 5.
Schematic of experimental setup and characterization of material (a) forward-osmosis process, (b) Dewatering process with thermal input, (c) multilayer material with the drawing layer for FO desalination, (d) releasing layer for fast water release, and (e) multi-layer design with gradual reduction of SA concentration along the water transport pathway (after [28]).

Author details

Alexander Fayer
Independent Researcher, Herzliya, Israel

*Address all correspondence to: alxpfayer@yahoo.com

References

[1] Im S-J, Jeong S, Jang A. Forward osmosis (FO)-reverse osmosis (RO) hybrid process incorporated with hollow fiber FO. Clean Water. 2021;**4**(1):51. DOI: 10.1038/s41545-021-00143-0

[2] Kim J, Hong S. Optimizing seawater reverse osmosis with internally staged design to improve product water quality and energy efficiency. Journal of Membrane Science. 2018;**568**:76-86. DOI: 10.1016/j.memsci.2018.09.046

[3] Im SJ, Jeong S, Jeong S, Jang A. Techno-economic evaluation of an element scale forward osmosis-reverse osmosis hybrid process for seawater desalination. Desalination. 2020;**476**. DOI: 10.1016/j.desal.2019.114240

[4] Chung T-S, Li X, Ong RC, Ge Q, Wang H, Gang H. Emerging forward osmosis (FO) technologies and challenges ahead for clean water and clean energy applications. Current Opinion in Chemical Engineering. 2012; **1**(3):246-257. DOI: 10.1016/j.coche.2012.07.004

[5] Wang J, Gao S, Tian J, Cui F, Shi W. Recent developments and future challenges of hydrogels as draw solutes in forward osmosis process. Watermark. 2020;**12**(3). DOI: 10.3390/w12030692

[6] Hackl EV, Khutoryanskiy VV, Tiguman GMB, Ermolina I. Evaluation of water properties in HEA–HEMA hydrogels swollen in aqueous-PEG solutions using thermoanalytical techniques. Journal of Thermal Analysis and Calorimetry. 2015;**121**:335-345. DOI: 10.1007/s10973-015-4446-y

[7] Lehmann S, Seiffert S, Richtering W. Spatially resolved tracer diffusion in complex responsive hydrogels. Journal of the American Chemical Society. 2012; **134**(38):15963-15969

[8] Zhao S. Osmotic pressure versus swelling pressure: comment on bifunctional polymer hydrogel layers as forward osmosis draw agents for continuous production of fresh water using solar energy. Environmental Science & Technology. 2014;**48**(7): 4212-4213. DOI: 10.1021/es5006994

[9] Horkay F, Magda J, Alcoutlabi M, Atzet S, Zarembinski T. Structural, mechanical and osmotic properties of njectable hyaluronan-based composite hydrogels. Polymer. 2010;**51**(19): 4424-4430. DOI: 10.1016/j.polymer.2010.06.027

[10] Fariba G, Farahani SV, Faraahani E. Theoretical description of hydrogel swelling: a review. Iranian Polymer Journal. 2010;**19**(5):375-398

[11] Kang MK, Huang R. Swell induced surface instability of confined hydrogel layers on substrates. Journal of the Mechanics and Physics of Solids. 2010; **58**(10):1582-1598. DOI: 10.1016/j.jmps.2010.07.008

[12] Beyhaghi S, Geoffroy S, Prat M, Krishna M. Pillai wicking and evaporation of liquids in porous wicks: a simple analytical approach to optimization of wick design. Transport Phenomena and Fluid Mechanics. 2014; **60**(5):1930-1940. DOI: 10.1002/aic.14353

[13] Hoffmann MR. Macroscopic Equations for Flow in Unsaturated Porous Media [Thesis]. Wageningen: Wageningen Universiteit; 2003. Available from: https://edepot.wur.nl

[14] Geisea M, Parkb H, Saglea AC, Freemana BD, McGrathc JE. Water

permeability and water/salt selectivity trade off in polymers for desalination. Journal of Membrane Science. 2011;**369** (1–2):130-138. DOI: 10.1016/j.memsci. 2010.11.054

[15] Carr EJ. Characteristic time scales for diffusion processes through layers and across interfaces. Physical Review E. 2018;**97**:042115. DOI: 10.1103/ PhysRevE.97.042115

[16] Kiil F. Molecular mechanisms of osmosis. The American Journal of Physiology. 1989;**2564**(Pt 2). DOI: 10.1152/ajpregu.1989.256.4.R801

[17] Kiil F. Mechanism of osmosis kidney international. Kidney International. 1982;**21**:303-308. DOI: 10.1038/ ki.1982.22

[18] Perry M. How forward osmosis (FO) performance is limited by concentration polarization. 2013. Available from: htps://www.forward osmosistech.com/how-forward-osmosis-performance-is-limited-by-concentra tion-polarization/

[19] Eyvaz M, Arslan S, İmer D, Yüksel E, Koyuncu İ. Forward osmosis membranes – a review: Part I. 2018. Available from: https://www.intechope n.com/chapters/59495. DOI: 10.5772/ intechopen.72287

[20] Fayer A. Continuous Cycle of Water Desalination Utilizing Hydrogel as Draw Agent. 2021. Available from: https:// www.scienceopen.com/hosted-docume nt?doi=10.14293/S2199-1006.1.SOR-. PPVRF7X.v1

[21] Soleimani F, Sadeghi M. Synthesis of pH-sensitive hydrogel based on starch-polyacrylate superabsorbent. Journal of Biomaterials and Nanobiotechnology. 2012;**3**:310-314. DOI: 10.4236/ jbnb.2012.322038

[22] Suzuki A, Ishii T, Maruyama Y. Optical switching in polymer gels. Journal of Applied Physics. 1996;**80**(1):1. DOI: 10.1063/1.362768

[23] Liu Q, Suo Z. Osmocapillary phase separation. Extreme Mechanics Letters. 2016;7:27-33. DOI: 10.1016/J. EML.2016.02.001

[24] Tauro F. Particle tracers and image analysis for surface flow observations. Wiley Interdisciplinary Reviews: Water. 2015;**3**(1). DOI: 10.1002/wat2.1116

[25] Hou Y, Ma S, Hao J, Lin C, Zhao J, Sui X. Construction and ion transport-related applications of the hydrogel-based membrane with 3D nanochannels. Polymers. 2022;**14**(19):4037. DOI: 10.3390/polym14194037

[26] Cheng P, Wang D, Schaaf P. A review on photothermal conversion of solar energy with nanomaterials and nanostructures: from fundamentals to applications advanced sustainable systems. 2022;**6**(9). DOI: 10.1002/ adsu.202200115

[27] Razmjou A, Liu Q, Simon GP, Wang H. Bifunctional polymer hydrogel layers as forward osmosis draw agents for continuous production of fresh water using solar energy. Environmental Science & Technology. 2013;**47**(22): 13160-13166. DOI: 10.1021/es403266y

[28] Zeng J, Cui S, Wang Q, Chen R. Multi-layer temperature-responsive hydrogel for forward-osmosis desalination with high permeable flux and fast water release. Desalination. 2019;**459**:105-113. DOI: 10.1016/j.desal. 2019.02.002

Chapter 6

Perspective Chapter: Technological Advances in Harnessing Energy from Renewable Sources for Water Production

Wafa Suwaileh, Rima Isaifan, Reza Rahighi, Amirmahmoud Bakhshayesh and Mohammad Ahmed

Abstract

Recently, different technologies such as desalination processes have been utilized to obtain fresh water from natural sources to develop good standards of life, flourish industrial activities, and enhance civilization. Hence, this book chapter aims to cover the fundamental aspects of harnessing energy from the sun or solar cells, covering the history of this topic as well as the new related policies. A discussion of the basics of solar cell devices, performance challenges, and long-term stability will follow. This chapter will also address state-of-the-art membrane-based desalination technologies in generating fresh water from various renewable sources such as solar, wind, wave, and geothermal.

Keywords: renewable energy, photovoltaics, water desalination, membrane technology, electrodialysis

1. Introduction

Water purification from industrial waste, potable water produced from natural sources, and filtration of contaminants (chemical or biological) from drinking water are the inevitable large-scale projects of our current decade. Given that approximately 80% of the world's population is at high risk of having inadequate access to clean water [1]. Fresh water is projected to become a scarce resource for many communities, creating a pressing need for new technological advancements in water purification for household systems. Additionally, the abundant low-grade water is suitable for producing economical and clean energy. Therefore, different technologies have been utilized in the last two decades to obtain fresh water, and there is a necessity to have a critical overview of the related processes and their corresponding mechanism toward enhancing and achieving higher yields.

In general, desalination is either referred to as thermal desalination or membrane-based processes. Desalination can be categorized into thermal process or distillation based on evaporation and condensation. In contrast, membrane-based processes use the membrane to separate salt from seawater or brackish water to generate fresh water [2].

The thermal desalination processes involve multi-stage flash distillation (MSF), multi-effect distillation (MED), humidification dehumidification (HDH), vapor compression distillation (VCD), adsorption desalination (AD), and freezing [2–4]. Among these processes, conventional MSF plants produce around 84% of the global desalination capacity, while MED plants generate about 3.5% of the world's desalted water [5]. This thermal process consumes more energy to remove salts from seawater than membrane-based techniques. The world's total desalination consumption capacity was predicted by 75.2 TWh of energy per year [6].

Membrane desalination techniques include reverse osmosis (RO), nano-filtration (NF), forward osmosis (FO), membrane distillation (MD), electrodialysis (ED), electrodialysis reversal (EDR), and capacitive deionization (CDI) [2–4, 7]. Their difference mainly depends on the size of contaminants in the solution that are separated or transported through the membrane, among which RO is known as the prevalent water desalination technology owing to its widespread availability in the market, cost-effectiveness, and efficiency in refining [3, 8, 9].

Every desalination process needs a different type of natural energy to operate its system. For example, wind, hydro, tidal, and wave are electricity producers, while solar, geothermal, and biomass are thermal and electrical energy producers [10]. Conventional multi-stage flash distillation (MSF) plants produce more than 80% of the global desalination capacity, while multi-effect distillation (MED) plants generate less than 4% of the world's desalted water [5]. This thermal process is more energy-consuming compared to membrane-based techniques. Regarding specific energy consumption, the RO process requires low energy input of only about 5 kWh/m^3 compared to MED and MSF (~ 20 kWh/m^3 and almost 25 kWh/m^3, respectively) [2]. As RO processes are energy-intensive, the high cost of water production will be alarming for billions of residents worldwide. The world's total desalination consumption capacity was predicted to be more than 75 TWh annually [6]. More progress over the last decade has emerged in harvesting natural energy-powered desalination systems to be cost-effective, which may appreciably reduce specific energy requirements and become beneficial, eco-friendly, and sustainable compared to the traditional desalination processes [11]. Herein, the most common natural energy, such as solar, wind, geothermal, and wave energy-combined desalination systems, shall be described in the following section.

As an infinite energy resource, solar energy has been employed for centuries and is considered a panacea for making mankind less dependent on fossil fuels. In fact, the last two decades witnessed almost a 300% increase in electricity capacity from renewable energy sources, including solar energy. The trend is expected to soar until 2030 with the advent of further wind turbines or solar panel installations by reviewing the different categories of solar cell technology made as photovoltaic devices to convert solar energy into electricity. FO process (derived by a gradient in osmotic pressure between low salinity and high salinity solution) can benefit from solar energy. Similarly, different low-cost PV systems can be integrated into normal RO processes to replace traditional fossil fuels with the natural sunlight energy source. Solar-driven MD process is another technique that can thermally separate unwanted ions from water. In this process, a porous hydrophobic membrane is utilized between hot and cold solutions to produce drinking water.

ED process can be integrated with the advances in solar cell technology to prevent further global warming and generate fresh water via solar-powered ED as a promising method that requires relatively low capital cost and specific energy consumption. However, it is limited to brackish water, and it needs further optimization of the experimental parameters (on large-scale operations) and be tested with seawater with different salinities. This book chapter also talks about the contribution of new techniques for water purification and even generation, e.g., via desalination and membrane technologies. The basics of other natural energy-driven desalination systems, such as marine wave, geothermal, and wind energies, are also discussed. This is due to the ever-increasing scarcity of reachable water and the dire need to shift energy harvesting from traditional grids to renewable energy sources. This chapter will, therefore, cover fundamental aspects of harnessing energy from the sun, wind, tides, and waves, followed by a discussion of the past and recent energy technologies and their feasibility. It also highlights the advantages and disadvantages of each harnessing system and the challenges faced earlier. Hence, this chapter will present state-of-the-art technological advances in harnessing energy from renewable sources to produce drinking water.

2. Solar cells

2.1 History and new policies

The world's first solar collector was built by Horace de Saussure in 1767 and was practically used in solar cookers in the 1830s [12]. Robert Stirling's system, invented in 1816, offered a solar thermal electric technology that concentrates the sun's thermal energy to produce power [13]. Edmond Becquerel's discovery of the photovoltaic (PV) effect in 1839 opened a new window in developing research on solar energy systems [14]. A few decades later, Augustin Mouchot registered an invention for solar-powered engines in the 1860s [15]. A decade later, Willoughby Smith discovered the photoconductivity of selenium working on underwater telegraph cables [16].

In 1883, Charles Fritts fabricated the first solar cell with sunlight-to-electricity efficiency as low as one percent by using selenium wafers coated with a thin layer of gold [17]. In 1888, Edward Weston received two granted patents (patent No. US389124 and No. US389425) focusing on transforming the sun's radiant energy into electrical energy. Meanwhile, Aleksandr Stoletov reported that solar cells based on the photoelectric effect create more power when exposed to ultraviolet light than visible light [18]. Clarence Kemp patented the first commercial solar water heater in 1891 (patent No. US451384A). Afterward, Melvin Severy received two patents on apparatus for mounting and operating thermopiles and apparatus for generating electricity by solar heat in 1894 (patent No. US527377A and US527379A, respectively).

William Coblentz received a patent in 1913 (patent No. 1077219A) for preparing a thermal generator that used light rays to generate an electric current. One year later, the existence of a barrier layer in photovoltaic devices was noted. Various research was in hand until semiconductor technology was born in Bell Laboratories in the 1950s (patent No. US2780765A), when scientists realized that silicon is more efficient for photovoltaic applications than selenium so that the produced solar cell showed an efficiency of around 6% [19]. The efficiency exceeded 10% in just 18 months [20]. The proficiency of their device was demonstrated using a solar-powered radio transmitter. This development was considered a key measure for filling the technology gap in solar energy harvesting, so *The New York Times* reported that the silicon solar cell "may mark the beginning of a new era, leading eventually to the realization of one of mankind's most cherished dreams–the harnessing of the almost limitless energy of the sun for the uses of civilization." In the same year, Texas Instruments Inc. filed a patent application on silicon p-n junctions (patent no. US2949498A), and Western Electric Co. began to sell commercial licenses for silicon photovoltaic technologies. The first commercial office building (i.e., the Bridgers-Paxton Building) using solar water heating was built in 1955. One year later, U.S. Signal Corps Laboratories focused on preparing solar panels for proposed orbiting Earth satellites, where they later developed n-on-p silicon photovoltaic cells. With this in mind, Vanguard I, Explorer III, Vanguard II, and Sputnik-3 have been launched with PV-powered systems on board as the accepted energy source for space applications and have remained so till today. In 1963, Sharp Corporation succeeded in producing practical silicon photovoltaic modules. Up to 1977, total PV manufacturing power was approximately 500 kilowatts, and ARCO Solar became the first company to produce more than 1 megawatt of photovoltaic modules in 1980 [21].

The National Renewable Energy Laboratory (NREL) was launched by the U.S. Department of Energy (DOE) in 1977 as a solar energy research institute dedicated to transforming energy through research, development, commercialization, and deployment of renewable energy and energy efficiency technologies. NREL is one of the best research-cell efficiency charts published since 1976 for a range of photovoltaic technologies on a standardized basis, confirmed by various independent test labs (e.g., NREL, AIST, JRC-ESTI, and Fraunhofer-ISE) [22, 23]. The chart includes solar cells within five families of semiconductors: multi-junction cells, single-junction gallium arsenide (GaAs) cells, crystalline silicon (Si) cells, thin-film technologies, and emerging photovoltaics, which are comprised of 28 subcategories, as summarized in **Table 1**. Thus far, multi-junction, single-junction gallium arsenide, and crystalline silicon cells have recorded the highest cell efficiencies, all of which take dominant positions in the market. Still, generally, these technologies are very costly today. Therefore, many researchers worldwide strive to develop thin-film technologies (e.g., CIGSs) and emerging technologies (e.g., perovskite structures), thanks mainly to their lower costs and facile fabrication processes.

During the past decades, the public and private sectors have made numerous energy policies to influence the global solar electricity market. Many local, national, and international projects have been carried out in this regard. Flat-Plate Solar Array (FSA) was a ten-year project run in 1975 to support developing PV modules with 10% efficiency, a 20-year lifetime, and a selling price of $0.50 per watt [24]. In 1978, the US Congress enacted the Public Utility Regulatory Policies Act (PURPA) in response to the energy crisis, with oil expected to rise to over $100 per barrel to stimulate alternative energy sources. PURPA obligated electric utilities to purchase renewable energy and is still driving renewable energy development so that it accounted for over 40% of the solar energy projects built in the US as of 2017.

The International Energy Agency Photovoltaic Power Systems Program (IEA PVPS), as one of the Technology Collaboration Programs (TCP) established in 1993, has recently reported that approximately 946 GW of PV power plants were producing electricity worldwide at the end of 2021, of which around 70% have been installed during the last five years [25]. Around 174 GW of new solar capacity was established in 2021, and that figure might rise to 260 GW in 2022. It is worth mentioning that 42 countries reached at least 1 GW of solar-based electricity in 2021. **Figure 1** illustrates the worldwide evolution of cumulative PV installations between 2011 and 2021. China continues to drive the global PV market, but the EU, USA, India, and Japan also play a vital role. EU ranked second in terms of newly installed solar capacity with 29 GW, after China with 55 GW.

China's new nationally determined contribution (NDC) pledges aim to hit the renewables target of 1.2 TW wind and solar power capacity by 2030. The Solar Energy

No.	Photovoltaic Technologies	Subcategories	Highest Confirmed Efficiencies (%)
1	Multi-junction Cells (2-terminal, monolithic)	Four-junction or more (concentrator)	47.1
		Three-junction (concentrator)	44.4
		Three-junction (non-concentrator)	39.5
		Four-junction or more (non-concentrator)	39.2
		Two-junction (concentrator)	35.5
		Two-junction (non-concentrator)	32.9
2	Single-Junction GaAs	Concentrator	30.8
		Thin-film crystal	29.1
		Single crystal	27.8
3	Crystalline Si Cells	Single crystal (concentrator)	27.6
		Silicon heterostructures (HIT)	26.7
		Single crystal (non-concentrator)	26.1
		Multicrystalline	23.3
		Thin-film crystal	21.2
4	Thin-Film Technologies	CIGS	23.4
		CIGS (concentrator)	23.3
		CdTe	22.1
		Amorphous Si: H (stabilized)	14.0
5	Emerging PV	Perovskite/Si tandem (monolithic)	31.3
		Perovskite cells	25.7
		Perovskite/CIGS tandem (monolithic)	24.2
		Organic cells	18.2
		Quantum dot cells (various types)	18.1
		Organic tandem cells	14.2
		Dye-sensitized cells	13.0
		Inorganic cells (CZTSSe)	13.0

Table 1.
NREL's best research-cell efficiency chart was updated on June 30, 2022. Adapted with permission from [22].

Industries Association (SEIA) has recently issued a roadmap chart to achieve the US solar manufacturing capacity of 10 GW in two years, 15 GW in three years, and 25 GW in five years on its path to 50 GW of annual production by 2030. Similarly, the European Commission increased its solar power capacity target to nearly 600 GW by 2030, with an interim target of 320 GW by 2025 [26]. Furthermore, a new solar target represents a 41% increase in ambitions compared to 420 GW under the EU's "Fit for 55" climate plan. Solar electricity generation capacity is expected to double every two years in the decade between 2020 and 2030 [25].

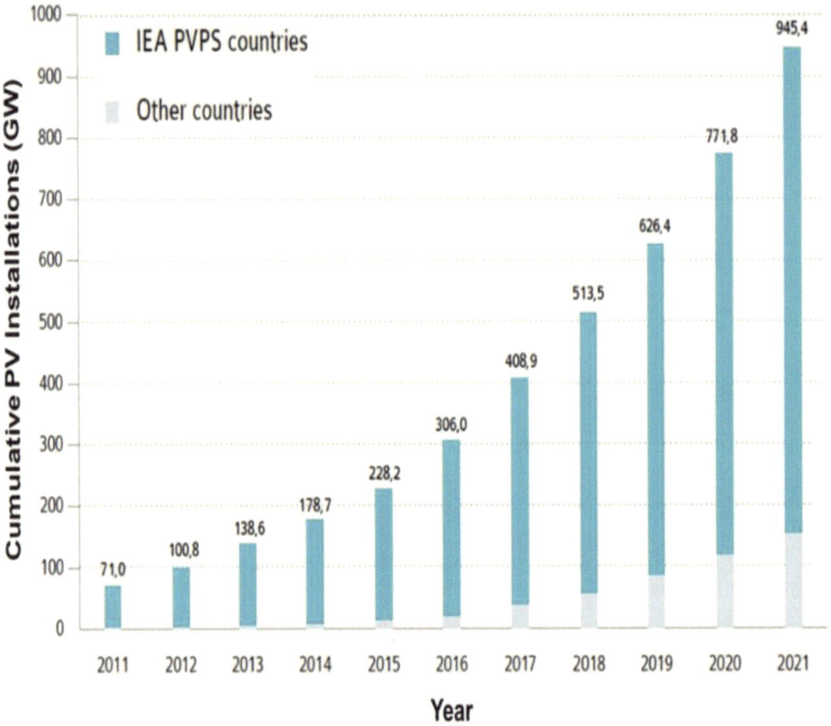

Figure 1.
Comparing the worldwide cumulative PV installations of IEA PVPS countries (Australia, Austria, Belgium, Canada, Chile, China, Denmark, Finland, France, Germany, Italy, Japan, Malaysia, Mexico, Morocco, the Netherlands, Norway, Portugal, South Africa, Spain, Sweden, Switzerland, Thailand, Turkey, and the United States of America) with other countries between 2011 and 2021. Adapted with permission from [25].

2.2 Device basics and performance challenges

Light-matter interaction can be harvested as green energy by converting sunlight into electricity through the photovoltaic effect. A silicon-based solar cell is essentially a diode with an n-type and p-type silicon configuration, which contains a level of impurities, improving the silicon potential to harvest energy from the sun and convert it into electricity. The p-type silicon is produced by adding the atoms having one less electron in their outer energy level than silicon (e.g., boron), creating an electron vacancy, namely a hole. The n-type silicon is made by adding atoms having one more electron in their outer level than that of silicon (e.g., phosphorus), creating an excess free electron not involved in bonding. A depletion zone is made around the junction of the layers, where free electrons can transfer from the n-type layer to the p-type layer to fill the holes. Under the radiation of photons, these electrons obtain sufficient energy to escape to the n-type layer and remake the holes in the p-type layer, producing a flow of electricity.

Poly-crystalline silicon solar cells, the most commercially available solar energy devices, suffer from restricted light-to-electricity conversion power. As the most efficient material for sunlight conversion, single-crystalline solar cells are made from very pure silicon and are employed for space applications. However, silicon-based solar panels are still reasonably expensive to practically change the quality of people's life. Rigidity and fragility are the other dominant Achilles' heels for transporting silicon panels. Besides, solar panels are expected to work for 25 years with a minimum of 80% of their original power, which has already been an eye-catching challenge to be addressed [27].

Over the past years, emerging technologies have played a vital role in enhancing the performance of generic silicon solar cell technologies and laid the foundations for disruptive solar cell technologies based on thin films and nanostructures. Thus far, a wide variety of first-generation solar cells have been commercially introduced, but many scientists from around the world are still working to develop more and more efficient next-generation solar cells; perovskite, dye-sensitized, quantum dot, polymer, copper indium gallium selenide (CIGS), copper zinc tin sulfide (SZTS), cadmium telluride (CdTe), and gallium arsenide (GaAs) solar cells as cases in point. Since the significant advances in solar technology have, directly or indirectly, their roots in nanotechnology, the future of solar power generation is in the hands of nanotechnology. A cursory glance at various available solar cell technologies is provided in the following sub-sections.

2.2.1 Perovskite solar cells

As an emerging solar cell family, perovskite-based ones, referring to a large group of chemicals with the general formula of ABX_3, have been the fastest to reach higher efficiencies. Organo-halide compounds based on Pb (i.e., B) and monovalent ions such as fluorine, chlorine, bromine, and iodine (i.e., X) have shown firm promise for the facile and low-cost production of solar cells. However, perovskite solar cells are still extensively studied due to stability challenges in sunlight, moisture, and heat exposure. The components of a typical perovskite solar cell are as follows [28]:

1. a light-absorbing layer made of perovskite compounds (e.g., $CH_3NH_3PbI_3$)

2. a transparent conductive oxide (TCO) substrate, which is a glass coated with fluorine-doped tin oxide (FTO)

3. a titanium dioxide (TiO_2) or silicon dioxide (SiO_2) blocking layer, which restricts the charge recombination process between the FTO and the light-absorbing layer

4. a mesoporous layer composed of TiO_2, Al_2O_3, ZnO, and ZrO_2 nanoparticles as a scaffold for the perovskite layer

5. a hole transport material (e.g., Spiro-OMeTAD)

6. a layer of gold, silver, or carbon as a cathode.

2.2.2 Dye-sensitized solar cells (DSSCs)

A mesoporous layer of 20-nm TiO_2 particles deposited on an FTO glass lies at the heart of a dye-sensitized solar cell. The dye molecules (e.g., N-719) introduced onto the surface of the TiO_2 nanoparticles are in charge of producing electron–hole pairs. Photons pass through the FTO glass and get absorbed by the dye molecules, which oxidize the dyes, providing excited electrons. Afterward, the excited electrons are moved to the molecules' lowest unoccupied molecular orbital (LOMO), leaving the holes behind in the highest occupied molecular orbital (HOMO) of the molecules.

The excited electrons diffuse to the TiO_2/FTO interface, finally reaching the FTO through the external circuit. The oxidized dye molecules can accept electrons from the electrolyte to be reduced to the ground state using I^- ions of the electrolyte, leading to the formation of I^{3-} ions. The remaining I^{3-} ions diffuse toward the cathode, where they are reduced to I^- ions again [29].

2.2.3 Quantum dot solar cells

As a substitute for dyes in DSSCs, quantum dots (QDs) create a new class of solar devices called quantum dot-sensitized solar cells, featuring high-density electron injection, and tuning the band gap [30]. A broad spectrum of QDs has thus far been introduced for use in quantum dot solar cells, such as lead sulfide (PbS), cadmium selenide (CdSe), and cadmium telluride (CdTe), opening a new window for developing Schottky, multi-junction, bulk heterojunction, and depleted heterojunction solar panels [31].

2.2.4 Polymer-based solar cells

Polymers deposited on flexible substrates (e.g., polyethylene terephthalate foils) empower researchers to build flexible solar panels. In this regard, a layer of indium tin oxide (ITO) as the anode; a layer of PEDOT: PSS serves as an electron blocking layer that helps transport holes to the anode; a layer of P3HT: PBCM, in which P3HT acts as the donor material and PBCM as the acceptor material; and an aluminum layer deposited as the back electrode are the main components [32].

2.2.5 Copper indium gallium selenide (CIGS) solar cells

As a p-type semiconductor having a thickness of around 2 mm, copper indium gallium selenide is commonly employed as the light-absorbing layer in CIGS solar cells. Additionally, a layer of molybdenum which is deposited on a piece of glass by sputtering, an n-type buffer layer commonly made of cadmium sulfide (CdS), a layer of intrinsic zinc oxide that protects the CdS and CIGS layers from sputtering damage while depositing the back electrode, and a layer of aluminum-doped ZnO which is deposited as the back electrode are the other parts of a typical CIGS solar device [33].

2.2.6 Copper zinc tin sulfide (CZTS) solar cells

CZTS and CIGS solar cells enjoy the exact mechanism and general structures. The difference is that the light-absorbing layer made of CZTS with the chemical formula of Cu_2ZnSnS_4 shows a lower level of toxicity than that of the CIGS layer. Besides, the abundance of CZTS's elements in nature compared to CIGS is a determining factor for drawing attention [34].

2.2.7 Cadmium telluride (CdTe) solar cells

A cadmium telluride layer as a p-type semiconductor is used as the light-adsorbing layer. A piece of ITO-coated glass as the front electrode, a layer of polycrystalline cadmium sulfide (n-type semiconductor) serves as a window layer, and an Al or Au layer deposited as the back electrode is the other vital layer making this family of solar cells [35].

2.2.8 Gallium arsenide (GaAs) solar cells

As mentioned in the previous section, GaAs solar panels are mainly used in the aerospace industry due to their high production cost. Monocrystalline layers of n-type and p-type GaAs are the main components of this family of panels, for which various thin film structures have so far been introduced. GaAs are commonly employed in single-junction solar cells (e.g., GaAs, InP, and InGaP), double-junctions (e.g., InGaP/GaAs), triple-junctions (e.g., InGaP/GaAs/Ge, InGaP/GaAs/InGaAs, and InGaP/GaAs/InGaAsNSb), and four-junctions [36].

2.2.9 Amorphous silicon (a-Si) solar cells

Compared to crystalline silicon solar panels, the principal privilege of a-Si cells is their potential for producing highly low-weight structures using such practical approaches as chemical vapor deposition. This family of cells comprises various p-i-n junctions using n-type and p-type silicon layers. A layer of intrinsic silicon, a layer of ITO, a layer of aluminum-doped ZnO, and a layer of silver as the back electrode are the principal layers of a typical a-Si solar cell [37].

3. Solar cell-driven desalination

Solar energy is the most widely abundant natural energy in the world and can be classified into solar thermal and photovoltaic [7]. The desalination system can supply one or both sources to run its process. The direct solar collecting approaches involved solar stills and humidification-dehumidification (HD) desalination in obtaining distilled water. Solar energy conversion into electricity or thermal energy is generally conducted through indirect methods like photophilic cells to supply MSF, MED, and RO desalination systems [38]. **Figure 2** presents various natural energy sources that can be used to power different desalination technologies. There has been a significant growth in research that explored the feasibility of solar energy-integrated desalination processes. An example of the solar thermal process is a solar collector. It captures the solar radiation and converts it to heat transferred into a liquid in the absorber.

The heat created from the collector can be transformed into both electric output and mechanical energy, which can be used directly to power the thermal desalination process or indirectly to run the membrane-based desalination system. Some crucial limitations are insufficient condensation and the recovery of latent heat from condensation, high cost, short lifetime, and energy storage problems, which hindered the commercialization and large-scale operation [6, 7, 9, 11, 38, 40].

3.1 Solar-driven RO process

The first solar PV-powered RO prototype was installed in the north of Mexico in 1979. This system exhibited a water production capacity of about 1.5 m^3/d and specific energy consumption of around 4 kwh/m^3 [41]. Another solar-powered RO desalination system was installed in Oman to desalinate brackish groundwater. The plant generates 5 m^3/day of freshwater during 5 h (of each day), and sometimes the maximum output becomes above 7.5 m^3/day. The average freshwater production cost was predicted at US$ 6.52/m^3 over the 20-year lifetime of the equipment. It is also recommended that solar tracking, adjusting the tilt angle, and continuous cleaning of the PV arrays may improve

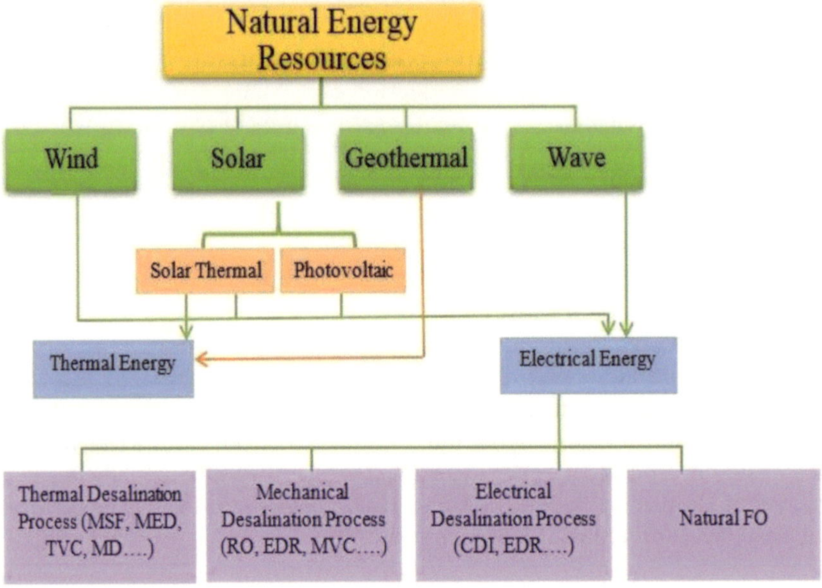

Figure 2.
A diagram showing the suggested integration of natural energy sources with different desalination processes. Reproduced from Sayed et al. [39].

the efficiency of the solar-powered desalination system. A theoretical study was conducted to determine the increase in the clean water flow rate for the yearly tilt, monthly tilt, and single- and double-axis tracking PV panels compared to the classical flat panel [42]. It was recorded that the annual permeate gain with the yearly optimal tilt was 10% and the monthly optimal tilt of the PV panel was 19%, respectively, concerning classical flat panel installation. When adjusting the PV orientation by adding a single- and double-axis tracker to the PV panel, the yearly permeate gain rose to 43% and 62%, respectively. However, the production of the RO process can be lowered due to the reduction in the radiation intensity during the low solar energy period or at night. Some issues are the cost of the PV cell, short experimental time, low lifetime, and reduced efficiency due to high average temperature, especially in GCC countries [2]. Thus, it is necessary to use a battery storage and cooling system [43].

A pilot plant PV-BWRO combined with an additional battery storage plant was constructed in Hartha Village, Jordan [44]. This system consisted of eight PV modules with 433 Wp, two batteries (230 Ah, 12 V), a softener as a pretreatment method, and an RO filtration system. It was designed to deliver 9.3–53 L/h from brackish water at 1700 mg/L over 24 hours, while the specific energy consumption reached 1.9 kWh/m^3 without using the softener. However, it was increased to 13.82 kWh/m^3 when coupling the softener to the desalination unit. The PV operated the high-pressure feed pump, data acquisition system, sensors, and solar regulator. The efficiency of the PV modules was good at about 12%. The system worked continuously during the year and generated enough fresh water for the citizens in the town. Depending on the percentage of water recovery and the system operating pressure, the range of freshwater permeate was between 9.3 L/h and 53 L/h [45]. Since the battery required frequent replacement, which increased the cost of water production, it is recommended to use a fuel cell storage device (FCs). When the FCs were coupled to the PV-RO unit, the water production approached 150 m^3/day [39]. It was proven that this hybrid system achieved an acceptable cost of electricity and was economically feasible. Compared to the battery, the FCs showed a longer lifetime, higher efficiency, and less ecological consequence as the carbon emission could be minimized to 71 t CO_2/year.

Rahimi et al. [46] considered different scenarios depending on the other solar panel installations into the grid for small-scale RO systems. The renewable RO unit was implemented in Bandar Abbas with high solar radiation and low cost of electricity. They assessed the impact of energy storage systems, ERDs, and membrane characteristics on the performance of the renewable RO system. It was concluded that the specific total capital cost was 1270 USD/(m^3/day) compared to that for the local market (1200 USD/(m^3/day)). When using the PV to deliver electricity to the power grid, the RO system achieved a unit product cost of about 1.11 US$/$m^3$ and a payback period of around 6 years. The workable membrane was SW30HRLE-440i, with a minimum water flux of 0.9 LMH/bar, and the lowest active area of 37.2 m^2. It could raise the energy consumption of the RO unit, which boosted the capacity of the PV panels and the RO unit productivity. Moreover, if the PV is used only to operate the RO unit, the RO system achieves the lowest net present value. Consequently, PV is a viable natural energy source for the RO process, which was proved to be economically viable and competitive with a traditional fuel source.

3.2 Solar-driven ED process

Solar-powered electrodialysis reversal is a promising alternative with lower capital cost, specific energy consumption, resistance to chlorine, and maximum recovery ratio [43, 47, 48]. The ED system involves many cell pairs full of saline water and separated by cation and anion exchange membranes [49]. Since the direct current polarity is supplied to the anode and cathode, the negative ions are transported through the anion exchange membrane. In contrast, the positive ions are transported through the cation exchange membrane. To alleviate the deposition of these ions in one of the compartments, a reversal DC polarity can be applied every 20 minutes. For instance, Al Madani et al. [50] tested a small-scale commercial-type electrodialysis stack driven by PV cells. The system's design included 24 cell pairs arranged in four hydraulic stages and two electrical stages. The feed solution contained synthetic NaCl with different salinity ranging from 1000 to 5000 ppm and groundwater with a salinity of 3300 ppm. The experiment was run at a temperature between 10 and 40°C while the product flow rate varied from 50 to 300 gallons/day. It was found that the quality of the diluted product was enhanced at a low product flow rate while increasing the temperature further improved the quality of the diluted product. The salt rejection was the highest, about 99% for the synthetic NaCl and 95% for the groundwater at a low product flow rate of 150 gal/day.

In India, a PV-EDR pilot plant was installed to treat brackish groundwater with salinity ranging from 3600 ppm to 350 ppm [51]. Relative to solar-powered RO process, the PV-EDR system consumed lower energy input per water unit (75% less at 1000 ppm, 50% at 2000 ppm, and 30% less at 3000 ppm). It is because the ED stack consumed a direct voltage at the anode and cathode and did not require DC/AC inversion and batteries. Meanwhile, the system possessed an excellent recovery ratio of around 92%, good tolerance toward chlorine, and low feed water changes.

Ortiz et al. [52] developed a computational model for a battery-less solar PV-powered ED system. This model was used to predict the number of PV modules required, the workable configuration, and the electrical consumption for the system under optimal experimental conditions like meteorological conditions, the volume of brackish water, the concentration of brackish water, and the flow rate. The theoretical data was closer to the experimental results. For instance, when increasing the brackish water feed salinity, the desalination time, energy consumption of PV-operated ED system, and drinking water production cost were grown remarkably. At the highest feed water conductivity of 7000 μm/cm, the electrical consumption, water production cost, and desalination time approached 130 kWh/m^3, 0.32 USD/m^3, and 105 minutes, respectively. On the other side, the PV-operated ED system is limited to brackish water and needs further experimental investigations on the large-scale operation and economic analysis. Ultimately, the PV-powered ED system should be widely investigated on large-scale operations to optimize the experimental parameters and the design for seawater with different compositions and different locations.

3.3 Solar-driven MD process

Membrane distillation is a thermal separation process using a porous hydrophobic membrane between hot and cold solutions [53, 54]. The separation occurred due to the temperature variation between the membrane surfaces, allowing vapors to pass through the membrane and then condense on a cold compartment. As a result, pure water was produced while the salt was rejected ultimately. The MD process is classified into air gap membrane distillation, sweeping gas distillation, direct contact membrane, and vacuum membrane distillation. Extensive theoretical and experimental research was performed worldwide to assess the viability of the PV-powered MD process for generating fresh water. Guillén-Burrieza et al. [55] evaluated the performance of two prototype solar collector combined AGMD processes using 1 and 35 g/L NaCl solutions under accurate operating parameters over 2 years at Plataforma Solar de Almeria (PSA) Spain. Prototype-1 was a compact individual design, while Prototype-2 was composed of three stages connected in series. They investigated the effect of various experimental conditions on drinking water production and its quality, thermal efficiency, and recovery ratio. There was negligible impact on the water quality produced with high purity (2–5 μS/cm) if the feed flow rate, temperature, or salt concentration increased. Based on the thermodynamic analysis, both prototypes worked as excellent heat exchangers, and there was no heat loss during the experiment. Prototype 2 had higher heat recovery and water product quality when the number of modules increased. The lowest specific thermal energy consumption was 294 kWh/m^3 for prototype 2 and 1805 kWh/m^3 for prototype 1. The highest water permeates corresponded to 5.09 $L/(h{\cdot}m^2)$, lower than the theoretical data or bench scale test. These findings indicated that the design and scaling up of the solar power MD model require further experiments to be commercialized.

Saffarini et al. [56] compared the water production cost between SP-MD systems using Direct Contact (DCMD), Air Gap (AGMD), and Vacuum (VMD) configurations to identify the correlation between many design and experimental conditions and the water production cost in the solar-driven membrane distillation process. According to a cost comparison, the water production cost varied in different MD models at a recovery ratio of 4.4%. It was 12.7 USD/m^3, 18.26 USD/m^3, and 16.02 USD/m^3 for DCMD, AGMD, and VMD, respectively. Although the DCMD with heat recovery machine showed high conductive losses from the feed inlet to the permeate outlet, it was the best cost-effective model. Regarding the influence of the design parameters of the module, solar collector efficiency, and operation parameters, the water production cost was decreased when the feed temperature, active membrane area, and solar module efficiency were increased while decreasing air gap width and feed channel depth for the AGMD model. Increasing the feed flow rate for a laminar flow regime caused high pumping power and, therefore, high water production costs. The water cost was reduced for a turbulent flow regime when lowering the flow rate. It can be concluded that the selection of the MD model and operating parameters will impact the final water production cost.

Mohan et al. [57] developed a new design of a solar thermal poly-generation (STP) pilot plant to produce chilled water for air conditioning using an absorption chiller, clean drinking water with MD process, and domestic hot water by heat recovery, as shown in **Figure 3**.

Four configurations were tested using seawater with a salinity of under weather conditions of the United Arab Emirates as follows: 1- solar cooling mode, 2- cogeneration of drinking water and domestic hot water, 3- cogeneration of cooling and desalination, and 4- tri-generation. An illustration of the solar-powered MD processes is depicted in **Figure 3**. The experimental results showed that 25 kW of chilled energy is harnessed with a capital operation cost of around 0.6 in solar cooling mode. The solar thermal tri-generation plant used 23% more useful energy compared to the solar cooling mode. The energy recovery was estimated by 75% at 50°C for extracting domestic hot water. It was reported that the tri-generation configuration with single-stage membrane module integration generated distilled water around 4 L/h, and it was increased to 80 L/h with the dual stage (cogeneration of cooling and desalination). Furthermore, the payback period was found to be 9.08 years, and cumulative net savings were regarded as $454,000, considering an inflation rate of 10% for constructing a rooftop. According to the financial evaluation, the payback period can be reduced by 18% and cumulative net saving by 10.7% if the land cost is not included.

Chen et al. [58] proposed a systematic two-stage design strategy for the solar-powered MD system considering the actual yearly radiation intensity of Taiwan.

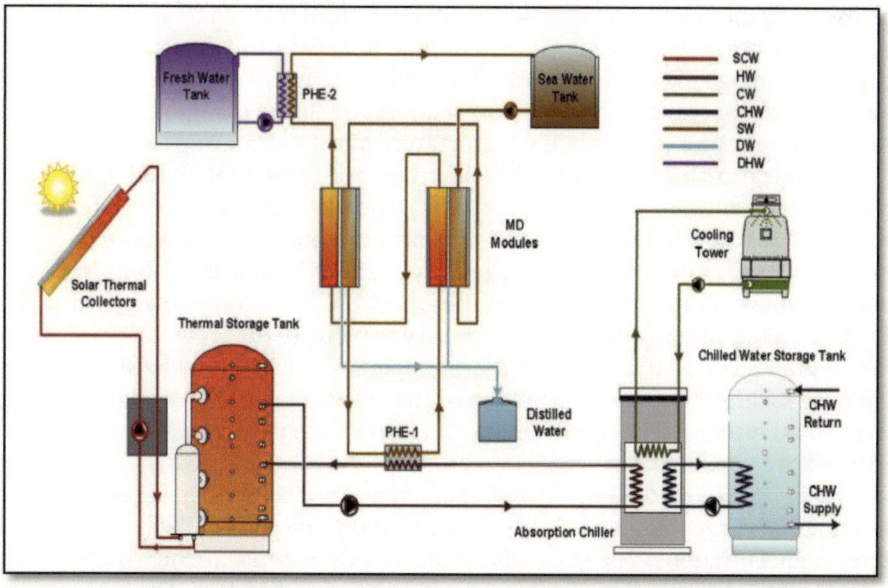

Figure 3.
A schematic drawing for the solar poly-generation system. Adapted with permission from Mohan et al. [57].

A dynamic simulation model developed on Aspen Custom Modeler's platform and a continuous operation control system were utilized to assess the performance of AGMD, DCMD, and VMD configurations. The first design worked under a fixed constant value of solar radiation intensity, and it was evaluated to determine the size of the compartments and the optimal experimental parameters. The second design was connected with the dynamic simulation model to change the flow rate automatically. It was revealed that the simulation results of the solar-powered MD process were in good correlation with the experimental literature results. When comparing all MD models, it was clear that the AGMD model needed the most extensive membrane distillation module, the DCMD model needed the most significant heat exchanger, and the VMD model required the most prominent solar collector. In terms of economic analysis, the water production cost of the optimal solar-powered MD system was \$2.71, 5.38, and 10.41 per m^3 of distilled water for AGMD, DCMD, and VMD models, respectively. Since the membrane was not expensive, its unit cost can be decreased from \$90/m^2 to \$36/m^2, resulting in a lower UPC of the optimal solar-driven AGMD system from \$2.71/m^3 to \$2.04/m^3. It should be noted that the price of membrane and solar is the main contributor to the capital cost and water production cost.

3.4 Solar-driven FO process

Forward osmosis is derived by the gradient in osmotic pressure between feed solution with low salinity and draw solution with high salinity [54]. This driving force allows the transport of the water molecules through a semi-permeable membrane to the draw solution side while blocking the transfer of salt ions. Forward osmosis was explored widely for water desalination. Still, the practical application of generating fresh water has been hindered due to some issues related to the membrane development, selection of draw solution, and low purity of water permeate. A novel process of integrating concentrated solar energy as the source of heat into the FO system was introduced by Razmjou et al. [59]. The tested draw solution was made of a water-absorptive layer to produce osmotic pressure, while the dewatering layer was at a lower critical solution temperature of 32°C to discharge the water being taken in during the experiment. The saline water was transported through the membrane and absorbed by the water-absorptive layer, which expanded and was recovered by a solar concentrator, as shown in **Figure 4**.

Layer concentrator increased the temperature of the dewatering layer to above its critical temperature leading to a high dewatering rate. Improving the energy input of the solar concentrator from 0.5 to 2 kW/m^2 yielded a rise in dewatering flux from 10 to 25 LMH. According to the thermodynamic evaluation, the minimum energy required for swollen dewatering hydrogel was 10.55 kWh/m^3 at a swelling ratio of around 5, a mass of dry hydrogel of about 1 g, and 100% recovery during hydrogel collapse. This high energy demand is necessary to increase the temperature of the hydrogel draw solution to above its critical temperature. Even though using more FO modules and a high swelling ratio promoted the FO water permeate and dewatering flux, it would increase the required energy to above the energy consumption of the pressure-driven desalination process.

Amjad et al. [60] studied the potential of a new draw solution made of potassium-functionalized carbon nanofibers dissolved in triethylene glycol (TEG) solution with different concentrations. The main advantages of this material are high osmotic pressure and excellent absorption of solar energy for the recovery step. The FO process was operated using a flat sheet FO membrane, synthetic 3.0 wt% NaCl solution as the feed solution, and 0.2 wt% concentration in 20 vol% TEG as the draw solution. The

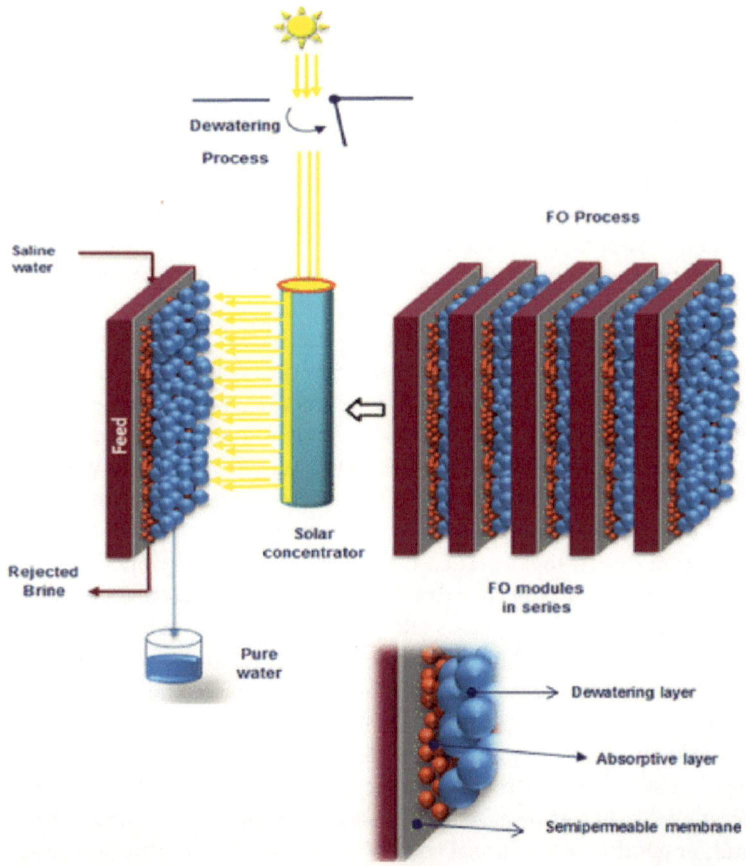

Figure 4.
A layout of solar concentrator combined FO process using bilayer thermo-responsive hydrogels as a draw solution and saline water as a feed solution for water generation. Adapted with permission from Razmjou et al. [59].

draw solution's recovery and freshwater extraction were carried out using simulated solar radiation flux. Based on the experimental results for the single FO process, the water flux was 13.3 LMH which was higher by 80% than that for 1 M NaCl as the draw solution against when DI water feed. When using the highest concentration of draw solution (0.2 wt% and 10 vol%), the reverse solute flux was as low as 0.031 g/L compared to that for 1 M NaCl as the draw solution against DI water feed. For the recovery step, the draw solution with the highest concentration exhibited improved photothermal efficiency by 105%, and the quality of the water permeate satisfied the quality standard of drinking water. Additionally, the efficiency of the draw solution was not changed after 5 cycles making it a favorable option for a prolonged solar-assisted FO-based desalination process.

Song et al. [61] investigated the possibility of extracting fresh water by combining the FO system with the photothermal evaporation method to regenerate the draw solution. A polyamide FO membrane was used for the FO process, and a photothermal polypyrrole nano-sponge (PPy/sponge) with large and interconnected pores to achieve high water evaporation was utilized for photothermal evaporation. The reduction of osmotic pressure due to the dilution of the draw solution in the single FO process was minimized by the interfacial water evaporation driven by solar energy.

A good balance was found between the water permeate in the individual FO system of about 11.2 m·L/h and water evaporation in the photothermal evaporation method of approximately 10 mL/h. As a result, a continuous water flux around 11.7 $L{\cdot}m^2/h^2$ and a low reverse salt flux around 3 $g/(m^2{\cdot}h^1)$ were obtained from the FO-EP system using NaCl as the draw solution with a salinity of 0.5 mol/L and light intensity of about 10 kW/m^2. This means that by controlling the concentration of the draw solution and light intensity, this system can produce continuous water permeate with low energy requirement and water production price.

A comprehensive statistical and experimental analysis was conducted to optimize the solar-powered FO process's experimental conditions and energy demand. Khayet et al. [62] developed a Monte Carlo simulation method to identify the optimum operating conditions and energy consumption of the solar thermal and PV-powered FO pilot plant. These experimental conditions were the water permeate flux, the reverse solute permeate flux, and the FO-specific performance index that involves the water and reverses solute permeate fluxes associated with the energy consumption. It used 35 g/L NaCl as the draw solution, DI water as the feed water, and a commercial spiral wound membrane module. It was found that 0.83 L/min feed flow rate, 0.31 L/min draw solution flow rate, and 32.65°C temperature were the optimum conditions and confirmed experimentally. The draw solution can be replenished using an optimized solar-powered reverse osmosis (RO) pilot plant with an optimum FO-specific performance index ranging from 25.79 to 0.62 L/g kWh under the same optimum conditions. This amount of energy represented only 14% of the total energy consumption for the solar-powered FO/RO hybrid process. In summary, to fully evaluate the system's economic viability, more studies are needed to achieve the commercial implementation of the solar-driven FO process for water generation from saline water or wastewater. Other researchers considered other natural sources like ocean waves, wind, or geothermal to boost the efficiency of the membrane-based renewable desalination system and minimize the water production cost.

4. Other natural energy-driven desalination

4.1 Wave energy-powered desalination system

Another promising energy carrier is wave energy, and it is possible to integrate it with desalination technology [63]. The most workable design of this hybrid system is a wave-activated body to pressurize seawater [64]. Then, this pressurized seawater transfers through a reverse osmosis membrane to produce drinking water. This kind of energy can be harnessed on coastal areas or islands with high wind potential. Regardless, the total theoretical energy that can be extracted from ocean waves was predicted by 8 x106 TWh/year, which is 100 folds the total hydroelectric power production of the world [65]. Wave energy coupled with desalination technology can be divided into two classes: directly or indirectly. In the first class, the movement of waves can be transformed into high pressure to operate the desalination system. In contrast, in the second class, electricity is exploited from the energy of the ocean waves [64]. Electricity is the dominant wave energy conversion in most previous studies. The first developed converter in the past was Delbuoy which consisted of a wave-driven buoy, linear pump, and an anchor system with the RO seawater desalination process to generate fresh water [66]. The electricity generated from the Delbuoy machine has been derived from the pumps connected to the ocean floor to transport the seawater to the RO desalination system. The systems are modular and capable of absorbing the required amount of radiation starting from 6 m^3 daily, depending on the plant's location. Eventually, this simple design does not need regular maintenance, lubrication, or chemicals and is inexpensive, which makes it an alternative option to traditional electric-driven desalination systems in developing countries. The first commercial wave energy-powered desalination plant (CETO) was constructed in Perth, Australia [67]. The plant utilized a buoy that was placed under seawater. The plant was composed of 3 units connected together and linked to the grid. The project started in 2014, and the plant operation ran for 12 months. The total wave energy was converted to the electricity of about 12,000 kWh over the demonstration cycle, which was exported to the grid. Some electricity-powered RO desalination plants generate around 150 m^3/day fresh water.

Recent work suggested using the Overtopping Breakwater for Wave Energy Conversion (OBREC) [68]. This work evaluated the performance of the OBREC-integrated RO desalination process for freshwater production on the Fenoarivo Atsinanana coast in Madagascar. In principle, this device converts the energy taken from the overtopping wave into electricity through heat turbines. This energy source contributed significantly to powering the RO desalination process. It was revealed that this renewable system could produce 964.3 m^3 per meter wave front of drinking water. In particular, 60 L per capita per day was provided to populations using a 500 m-long OBREC breakwater connected to the desalination plant. It could be used due to many advantages like low installation cost, minimum environmental impact, and high freshwater capacity.

Extensive research has explored the potentiality of harvesting wave energy around Grand Canary Island [69–71]. A successful wave energy/desalination project in the north of Gran Canaria Island was investigated by Prieto et al. [69]. Not only the system design was considered, but other factors like the identification of the installation area for the harnessing of wave energy, the type of the converter, its price, its efficiency for generating power, and commercialization are essential. In this case study, the location was selected based on the availability of many desalination plants in the strip of the coast between Punta del Camello and Punta de Guanarteme. This area is suitable for extracting wave energy to be supplied to the desalination plant, and there was no potential conflict. The annual average wave energy approached 20 kW/m, and a clear seasonal pattern ranged from 27 kW/m in winter to 12 kW/m in summer.

However, the intermediate values during the transitional seasons were regarded as 22 kW/m in autumn and 17 kW/m in spring. Hence, this wave energy is still acceptable for deriving the desalination plant. Also, the environmental impact is insignificantly associated with a great chance of reducing fossil fuel consumption. Another renewable desalination plant project followed the experience in the north of Gran Canaria, Spain. Cabrera et al. [71] developed a model for designing and operating wave-combined seawater RO desalination plants. The simulation findings showed that this hybrid renewable system could produce an average of $1.51 \cdot 10^5$ m^3/year of fresh water and supply around 1370 citizens or the agricultural water demand of 37 ha.

On the other hand, the specific cost of this system is high because this hybrid technology was explored in a few research projects. Another aspect is the number of converter devices used in the desalination plant. In this study, only one converter device was connected to the desalination plant but installing more converters with competitive prices in the market would eventually decrease the capital cost. Also, if an energy recovery device was replaced with a conservative specific energy consumption of 4 kWh/m^3, the water production cost could be minimized.

4.2 Geothermal energy-powered desalination process

In recent years, one of the most researched modes of integration between the natural energy source and desalination system is using geothermal energy to obtain fresh water [72]. Geothermal energy is the heat generated in the earth's underground and transferred to the water in contact with it [11, 73]. The significant advantages of this heat are that it is naturally available, continuous, and stable, with no need for energy storage. The hot water can be utilized to power a turbine and generate electricity [11, 40]. This output electricity can be used to operate different desalination plants. A previous study revealed that geothermal power is a beneficial natural energy source that is accessible every day of the year during the day and night which can provide 8% of the total global electricity [73]. The electricity obtained from geothermal is enough to supply 17% of the world's population. At the same time, citizens in Africa, Central/South America, and the Pacific can get all electricity from this clean energy. Karytsas et al. [74] stated that the geothermal energy extracted from the upper 2 Km of the hot rocks in Milos Island, Greece, can supply a 260 MWe geothermal power plant. The thermal energy harnessed from shallow with low enthalpy (<100°C) could run a water desalination unit providing 75–80 m^3/h of drinking water and turn an Organic Rankine Cycle turbine (ORC) with an installed capacity of 470 kW. This renewable energy-driven seawater desalination plant comprised hot seawater (80,000 ppm) from the deep well, which was used to operate the ORC turbine to produce electricity. Also, this hot seawater was passed through the MED as a boiling process to desalinate seawater. The MED consists of several stages, and the water is heated by steam in tubes in each stage. As a result, the evaporated water can be used as drinking water while the remaining salt water is transported to the next stage and becomes hot and evaporated. The vapors were condensed to collect fresh water. The energy generated from the previous stage was reused in the following stage. Subsequently, the population can be supplied with sufficient drinking water from a cheap, clean energy source (i.e., 1.5 € per m^3), ensuring water availability and food security. The poor efficiency of binary cycle systems when using the ORC and the Kalina cycle is a critical issue. In particular, the hot geothermal water with a temperature between 80 and 95°C was usually transferred to the desalination process providing low efficiency of around 5.5–8.5% [72].

An interesting example is given by Kaczmarczyk et al. [72], who proposed using low-enthalpy geothermal water to produce electricity for the desalination system and to supply the desalination process with a source of wastewater. This hot wastewater supplied to the RO desalination plant had low viscosity, allowing a large amount of fresh water. Also, it can be treated using other desalination technologies such as MD, MED, or MSF. The modeling results predicted that if the Kalina cycle is used for geothermal water (100 kg/s) at a temperature of 95°C and an ammonia-water mixture (89% ammonia and 11% water), a considerable amount of electricity can be exploited for about 6489 MWh/year. This electricity is sufficient to operate a geothermal energy-integrated brackish water desalination plant providing 3933 m^3 of daily drinking water. It is important to note that this low-temperature geothermal water has been proven to be an innovative choice for the water desalination process leading to sustainable development in water resources management.

In other words, a combination of solar and geothermal energies provided an attractive option to run the desalination plant aiming to accelerate electricity, improve the efficiency of the ORC module, and lower the extraction of geothermal sources [75–77]. In Calise et al. study [75], the solar power was derived from the ORC, and the geothermal power was supplied to the MED process for drinking water production.

In addition, geothermal energy was also used to power the Thermal Recovery Subsystem (TRS) connected to a cooling or heating device. Accordingly, generating electricity of about 4.60×10^3 MWh/year was possible, and heat recovery for heating and cooling was regarded as 8.31×10^4 MWh/year.

A significant breakthrough in the renewable desalination process is using a solar still as a humidifier and a ground heat exchanger as a condenser connected to an underground tube containing humid air [77]. The condensation of the humid air occurred in the condenser, and the resultant clean water was collected in the drainage pipe. A description of the system is shown in **Figure 5**. This freshwater could be used for irrigation or as drinking water. Since hot air with high temperature flows out continuously from the heat exchanger, it can be employed for cooling or heating purposes. Considering the theoretical analysis, it was concluded that the generation of drinking water was increased by 30.35% due to using a solar collector.

However, it was very low, around 4.45%, when using a solar air collector. Another suggestion is extracting the hot air flowing from the underground heat exchanger as its temperature remained high through the year at a depth of a meter. This can be accounted for by the fact that this hot air can provide extra energy and boost the performance of the solar photovoltaic and heating, ventilation, and air conditioning (HVAC) cycles. In this framework, the geothermal energy-powered desalination plant is economically profitable and environmentally reasonable for drinking water and irrigation water production in countries facing a freshwater crisis.

4.3 Wind energy-powered desalination system

The combination of wind energy and the desalination system is promising to produce fresh water at competitive cost and sustainability [78]. Compared with solar energy, the wind source is widely available in coastal and mountain areas and can supply power to desalination facilities without interruption [79]. One of the benefits is that the energy storage device is unessential. And its simple design, ease of installation, and limited maintenance price make it technically and economically competitive among other renewable energy sources. It could compete with other renewable systems in terms of a minimum environmental impact of about 75%, especially in locations with abundant wind power [78, 79]. The market's most popular wind power machine is the wind turbine, which is commonly coupled to the desalination plant for drinking water production. An exciting project on the combination of wind energy and RO desalination process was implemented in Drenec Island, France, in 1990. A wind turbine powered the seawater RO desalination process; rated at 10 kW. A wind turbine with a 5-kW can deliver energy of about 1500 Ah, 24 V to many batteries, and 1.8 kW to the RO desalination process [80].

Numerous types of research have been conducted on wind energy-coupled RO desalination systems such as AERODESA, SDAWES, and AEROGEDESA [80]. Work by Serrano-Tovar et al. [81] proposed using a wind farm-powered seawater RO desalination plant installed by Soslaires Canarias S.L. in 2002. Since the desalination plant cannot get all the necessary electricity capacity from the wind farm, part of the generated electricity from the wind farm is delivered to the electricity grid of Gran Canaria Island, and some electricity powers the desalination plant from the grid. The desalination plant and water pumping are operated by an electrical grid associated with an onshore wind power farm (2.64 MW). It should be noted that this desalination plant consumed an electrical grid when the wind power deteriorated. The seawater desalination plant is based on the RO process with wind electricity-derived water

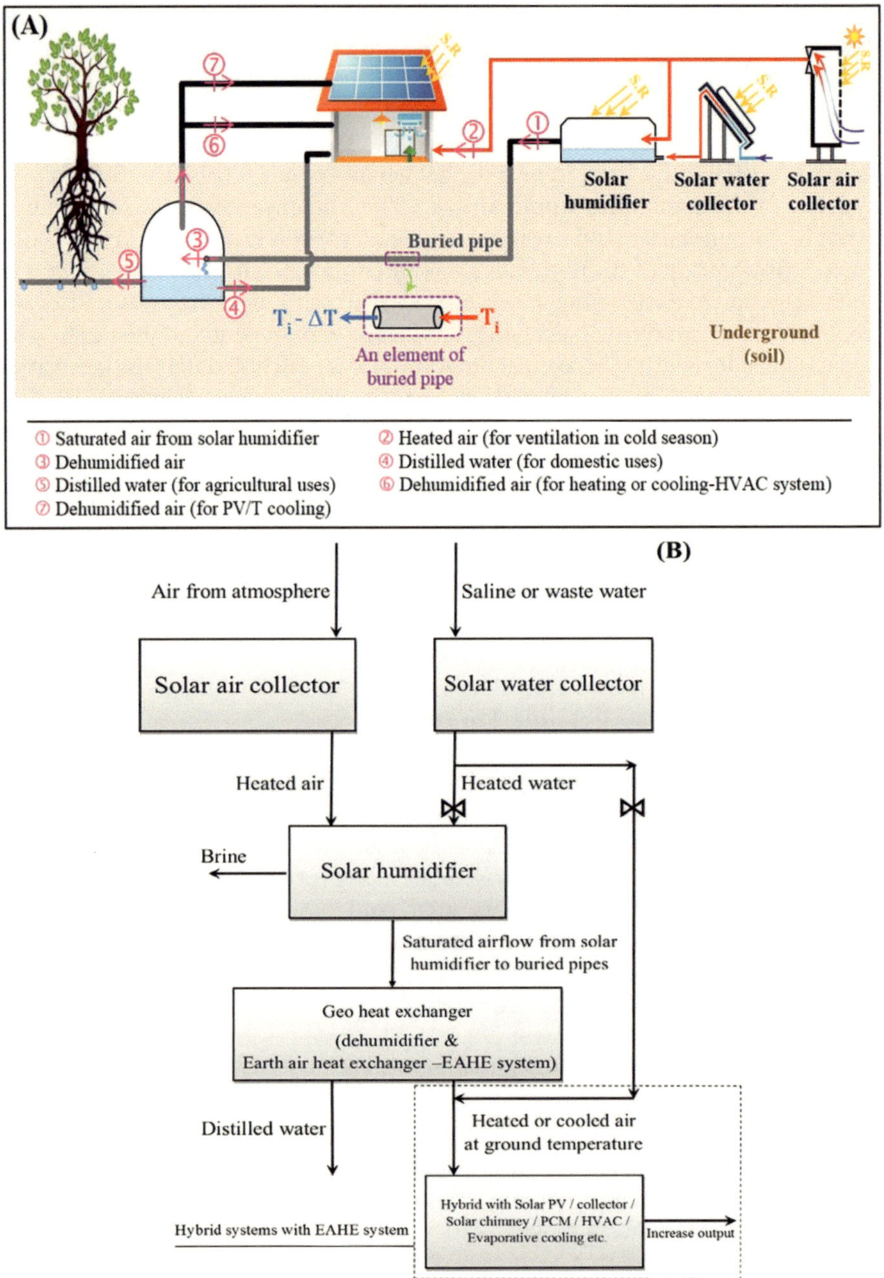

Figure 5.
A schematic diagram of (A) the geothermal energy-driven desalination system containing a solar desalinator as a humidifier and a ground heat exchanger as a condenser (B) a flowchart of the system's compartments linked together. Adapted with permission from Okati et al. [77].

capacity of around 5000 m^3/day. This product water was supplied to the agricultural land for irrigating 230 ha of fresh vegetables and fruits. In the study presented here, the optimum design of renewable energy combined desalination process, the product water capacity, the final efficiency, and the capital cost remained a real challenge. Besides, when an electrical grid powers the desalination plant, the performance of the combined system can be influenced by variations in power and sometimes cutoff caused by wind energy. Consequently, the desalination plant required a storage device like a battery, diesel generator, or flywheels to prevent power interruption [82].

Accordingly, Melian-Martel et al. [83] recommended deriving the entire water cycle on the Island involving drawing out groundwater, seawater desalination, water pumping, and distribution with abundant wind energy. The desalination process requires about 18% of the total electricity of the Island while producing groundwater, distribution, and storage processes require 17%. To determine the suitable renewable configuration without an energy storage device, the author suggested two different strategies: utilizing the current water supply network derived from seawater and groundwater, which was decentralized and using the wind-powered large-scale seawater RO desalination plant associated with a centralized water storage system. The main findings were that the design based on the first strategy yielded a growth of wind integration of around 22%, but it consumed a tremendous amount of groundwater of about 45% of total water extraction for irrigation. Conversely, the design based on the second strategy produced sufficient water capacity to meet the water requirement for the Island. In this case, the annual wind integration was increased by around 84%, achieving a higher contribution from natural energy. This means that significant potential exists in the sustainable management of aquifer water when combining wind energy-driven SWRO desalination plants with a centralized storage system. Considering the capital cost, additional storage devices are expensive, raising the water production cost.

5. Conclusions

The fast climate change and the ramifications of being on the verge of 1.9°C of global warming include longer warm seasons that insinuate shorter water resources. The great worldwide demand for potable water makes desalination inevitable. Some examples of RO desalination plants have recorded almost 150 m^3 of freshwater production daily. While RO is known as the most widespread membrane-based desalination process due to its superiority over thermal process (i.e., low-cost, easy-access/implement, and high efficiency), a combination of electrochemical schemes with electrodialysis reversal or capacitive deionization can even provide a higher rejection capability toward higher quality potable water. In general, the energy required for desalination processes can be supplied from natural wind, ocean waves, solar light, or geothermal, as discussed in this chapter.

Having discussed the basics of solar cell device fabrication, reviewing their state-of-the-art categories (such as multi-junction, thin-film crystals, silicon heterostructures, dye-sensitized, organic or inorganic, CIGS, perovskite/Si tandem, and quantum dot-based cells), and the horizon ahead for the future of extremely low-price roll-to-roll solar panels, one can foresee the perspective of next-generation solar-driven electrolyzers that can generate potable water while charging lithium- or even sodium-based battery of an automobile. In fact, freshwater can be a side-product of a fuel cell, while the future of large-scale high-pressure-output electrolyzer systems can be a source of energy for water desalination methods.

Conflict of interest

The authors declare no conflict of interest.

Author details

Wafa Suwaileh[1]*, Rima Isaifan[2], Reza Rahighi[3], Amirmahmoud Bakhshayesh[4] and Mohammad Ahmed[5]

1 Chemical Engineering Program, Texas A&M University at Qatar, Doha, Qatar

2 Division of Sustainable Development (DSD), College of Science and Engineering, Hamad Bin Khalifa University, Doha, Qatar

3 SKKU Advanced Institute of Nano-Technology (SAINT), Sungkyunkwan University, Seoul, South Korea

4 School of Metallurgy and Materials Engineering, Iran University of Science and Technology (IUST), Tehran, Iran

5 School of Materials Science and Engineering, Gwangju Institute of Science and Technology (GIST), Gwangju, South Korea

*Address all correspondence to: wafa.suwaileh@qatar.tamu.edu

References

[1] Tong W, Forster M, Dionigi F, et al. Electrolysis of low-grade and saline surface water. Nat Energy. 2020;**5**:367-377

[2] Abdelkareem MA, Assad MEH, Sayed ET, Soudan B. Recent progress in the use of renewable energy sources to power water desalination plants. Desalination. 2018;**435**:97-113

[3] Fornarelli R, Shahnia F, Anda M, Bahri PA, Ho G. Selecting an economically suitable and sustainable solution for a renewable energy-powered water desalination system: A rural Australian case study. Desalination. 2018;**435**:128-139

[4] Ahmadi E, McLellan B, Mohammadi-Ivatloo B, Tezuka T. The role of renewable energy resources in sustainability of water desalination as a potential fresh-water source: An updated review. Sustainability. 2020;**12**:1-31

[5] Shatat M, Riffat SB. Water desalination technologies utilizing conventional and renewable energy sources. International Journal of Low-Carbon Technologies. 2014;**9**:1-19

[6] Ali A, Tufa RA, Macedonio F, Curcio E, Drioli E. Membrane technology in renewable-energy-driven desalination. Renewable and Sustainable Energy Reviews. 2018;**81**:1-21

[7] Shokri MSF. A sustainable approach in water desalination with the integration of renewable energy sources: Environmental engineering challenges and perspectives. Environmental Advances. 2022;**9**:1-19

[8] Ghaffour N, Bundschuh J, Mahmoudi H, Goosen MFA. Renewable energy-driven desalination technologies: A comprehensive review on challenges and potential applications of integrated systems. Desalination. 2015;**356**:94-114

[9] Li S, Cai Y-H, Schäfer AI, Richards BS. Renewable energy powered membrane technology: A review of the reliability of photovoltaic-powered membrane system components for brackish water desalination. Applied Energy. 2019;**253**:1-23

[10] Curto D, Franzitta V, Guercio A. A review of the water desalination technologies. Applied Sciences. 2021;**11**:1-36

[11] Ghazi ZM, Rizvi SWF, Shahid WM, Abdulhameed AM, Saleem H, Zaidi SJ. An overview of water desalination systems integrated with renewable energy sources. Desalination. 2022;**542**:1-30

[12] Yaffe LF. Solar Cooking for Home and Camp. Mechanicsburg, PA, USA: Stackpole Books; 2007

[13] Kumari A. Solar stirling power generation. Journal of Refrigeration, Air Conditioning, Heating and Ventilation. 2020;**7**:7

[14] Williams R. Becquerel photovoltaic effect in binary compounds. The Journal of Chemical Physics. 1960;**32**:1505-1514

[15] Kryza FT. The Power of Light: The Epic Story of Man's Quest to Harness the Sun. New York: McGraw-Hill; 2003. p. 299

[16] Effect of light on selenium during the passage of an electric current. Nature. 1873;**7**:303

[17] Sharma R, Sharma A, Agarwal S, Dhaka MS. Stability and efficiency issues, solutions and advancements in perovskite solar cells: A review. Solar Energy. 2022;**244**:516-535

[18] Stoletow A. Sur une sorte de courants electriques provoques par les rayons ultraviolets. Comptes Rendus. 1888;**CVI**:1149

[19] Bell Laboratories Record. Western Electric Announces New Vice Presidents. 1954. pp. 436

[20] Bell Laboratories Record. Efficiency of Bell Solar Battery Almost Doubled. 1955. pp. 166

[21] Arnett JC, Schaffer LA, Rumberg JP, Tolbert REL. Design, installation and performance of the ARCO solar one-megawatt power plant. In: Fifth E.C. Photovoltaic Solar Energy Conference Proceedings, Kavouri (Athens). Greece: Smantic Scholar; 1983. pp. 314-320

[22] Research P. NREL Transforming Energy. National Renewable Energy Laboratory;

[23] Bulumulla C, Gamage PL, Miller JT, Kularatne RN, Biewer MC, Stefan MC. Pyrrole-containing semiconducting materials: Synthesis and applications in organic photovoltaics and organic field-effect transistors. ACS Applied Materials & Interfaces. 2020;**12**:32209-32232

[24] Christensen E. Flat-Plate Solar Array Project. Y.o.P., Department of Energy, United States; 1985

[25] Applications, T.I.P., REPORT IEA PVPS T1-43. Photovoltaic power systems technology collaboration programme, 2022

[26] 2019, R.P.G.C.i., Internation Eenewable Energy Agency, Abu Dhabi. 2020. Available from: https://www.irena.org/publications/2020/Jun/Renewable-Power-Costs-in-2019

[27] Fraas LP. Solar cells and their applications. 2nd edition. Hoboken, New Jersey: John Wiley & Sons, Inc.; 2010

[28] Correa-Baena JP, Abate A, Saliba M, Tress W, Jacobsson TJ, Grätzelc M, et al. The rapid evolution of highly efficient perovskite solar cells. Energy & Environmental Science. 2017;**10**:710-727

[29] B. O'Regan, M.G., A low-cost, high-efficiency solar cell based on dye-sensitized colloidal TiO2 films. Nature, 1991. 353: p. 737-740

[30] Liang M, Zhang J, Ramalingam K, Wei Q, Hui KS, Aung SH, et al. Stable and efficient self-sustained photoelectrochemical desalination based on CdS QDs/BiVO4 heterostructure. Chemical Engineering Journal. 2022;**429**:1-10

[31] Zhao Q, Hazarika A, Chen X, Harvey SP, Larson BW, Teeter GR, et al. High efficiency perovskite quantum dot solar cells with charge separating heterostructure. Nature Communications. 2019;**10**:1-8

[32] Nelson J. Polymer: fullerene bulk heterojunction solar cells. Materials Today. 2011;**14**:462-470

[33] J. Ramanujam, U.P.S., Copper indium gallium selenide based solar cells. Energy & Environmental Science, 2017. 10: p. 1306-1319

[34] K. Ramasamy, M.A.M.P.O.B., Routes to copper zinc tin sulfide Cu2ZnSnS4 a potential material for solar cells. Chemical Communications, 2012. 48: p. 5703-5714

[35] Lamb DA, Irvine SJC, Baker MA, Underwood CI, Mardhani S. Thin film cadmium telluride solar cells on ultra-thin glass in low earth orbit—3 years of performance data on the AlSat-1N CubeSat mission. Progress in Photovoltaics: Research and Applications. 2021;**29**:1000-1007

[36] Moon S, Kim K, Kim Y, Heo J, Lee J. Highly efficient single-junction GaAs thin-film solar cell on flexible substrate. Scientific Reports. 2016;**6**:1-6

[37] Stuckelberger M, Biron R, Wyrsch N, Haug F-J, Ballif C. Progress in solar cells from hydrogenated amorphous silicon. Renewable and Sustainable Energy Reviews. 2017;**76**:1497-1523

[38] Khan MAM, Rehman S, Al-Sulaiman FA. A hybrid renewable energy system as a potential energy source for water desalination using reverse osmosis: A review. Renewable and Sustainable Energy Reviews. 2018;**97**:456-477

[39] Sayed ET, Olabi AG, Elsaid K, Al Radi M, Alqadi R, Abdelkareem MA. Recent progress in renewable energy based-desalination in the Middle East and North Africa MENA region. Journal of Advanced Research. 2022:1-32

[40] Eltawil MA, Zhengming Z, Yuan L. A review of renewable energy technologies integrated with desalination systems. Renewable and Sustainable Energy Reviews. 2009;**13**:2245-2262

[41] Petersen G, Fries S, Mohn J, Müller A. Wind and solar powered reverse osmosis desalination units – description of two demonstration projects. Desalination. 1979;**31**:501-509

[42] Ahmad N, Sheikh AK, Gandhidasan P, Elshafie M. Modeling, simulation and performance evaluation of a community scale PVRO water desalination system operated by fixed and tracking PV panels: A case study for Dhahran city, Saudi Arabia. Renewable Energy. 2015;**75**:1-15

[43] Mohamed ES, Papadakis G, Mathioulakis E, Belessiotis V. A direct coupled photovoltaic seawater reverse osmosis desalination system toward battery based systems - a technical and economical experimental comparative study. Desalination. 2008;**221**:17-22

[44] Qiblawey H, Banat F, Al-Nasser Q. Performance of reverse osmosis pilot plant powered by Photovoltaic in Jordan. Renewable Energy. 2011;**36**:52-60

[45] Eltawil MA, Zhao Z, Yuan L. Renewable energy powered desalination systems: Technologies and economics-state of the art. In: Twelfth International Water Technology Conference, IWTC12 2008. Alexandria, Egypt; 2008. pp. 1099-1136

[46] Rahimi B, Shirvani H, Alamolhoda AA, Farhadi F, Karimi M. A feasibility study of solar-powered reverse osmosis processes. Desalination. 2021;**500**:1-21

[47] Teo HG, Lee PS, Hawlader MNA. An active cooling system for photovoltaic modules. Applied Energy. 2012;**90**:309-315

[48] Ramalingam K, Wei Q, Chen F, Shen K, Liang M, Dai J, et al. Achieving high-quality freshwater from a self-sustainable integrated solar redox-flow desalination device. Small. Jul 2021;**17**(30):e2100490. DOI: 10.1002/smll.202100490. Epub 2021 Jun 23. PMID: 34160139

[49] H. Sharon, K.S.R., A review of solar energy driven desalination technologies. Renewable and Sustainable Energy Reviews, 2015. 41: p. 1080-1118.

[50] AlMadani HMN. Water desalination by solar powered electrodialysis process. Renewable Energy. 2003;**28**:1915-1924

[51] N. C. Wright, A.G.W.V., Justification for community-scale

photovoltaic-powered electrodialysis desalination systems for inland rural villages in India. Desalination, 2014. 352: p. 82-91

[52] Ortiz JM, Expósito E, Gallud F, García-García V, Montiel V, Aldaz A. Photovoltaic electrodialysis system for brackish water desalination: Modeling of global process. Journal of Membrane Science. 2006;**274**:138-149

[53] Khayet M. Solar desalination by membrane distillation: Dispersion in energy consumption analysis and water production costs (a review). Desalination. 2013;**308**:89-101

[54] Suwaileh W, Johnson D, Jones D, Hilal N. An integrated fertilizer driven forward osmosis- renewables powered membrane distillation system for brackish water desalination: A combined experimental and theoretical approach. Desalination. 2019;**471**:1-17

[55] Guillen-Burrieza E, Zaragoza G, Miralles-Cuevas S, Blanco J. Experimental evaluation of two pilot-scale membrane distillation modules used for solar desalination. Journal of Membrane Science. 2012;**409-410**:264-275

[56] Saffarini RB, Summers EK, Arafat HA, Lienhard JH. Economic evaluation of stand-alone solar powered membrane distillation systems. Desalination. 2012;**299**:55-62

[57] Mohan G, Kumar NTU, Pokhrel MK, Martin A. Experimental investigation of a novel solar thermal polygeneration plant in United Arab Emirates. Renewable Energy. 2016;**91**:361-373

[58] Chen Y, Hung H-G, Ho C-D, Chang H. Economic design of solar-driven membrane distillation systems for desalination. Membranes. 2021;**11**:1-20

[59] Razmjou A, Liu Q, Simon GP, Wang H. Bifunctional polymer hydrogel layers as forward osmosis draw agents for continuous production of fresh water using solar energy. Environmental Science & Technology. 2013;**47**:13160-13166

[60] Amjad M, Gardy J, Hassanpour A, Wen D. Novel draw solution for forward osmosis based solar desalination. Applied Energy. 2018;**230**:220-231

[61] Song X, Dong W, Zhang Y, Abdel-Ghafar HM, Toghan A. Coupling solar-driven interfacial evaporation with forward osmosis for continuous water treatment. Exploration. 2022;**2**:1-8

[62] Khayet M, Sanmartino JA, Essalhi M, García-Payo MC, Hilal N. Modeling and optimization of a solar forward osmosis pilot plant by response surface methodology. Solar Energy. 2016;**137**:290-302

[63] Franzitta V, Curto D, Milone D, Viola A. The desalination process driven by wave energy: A challenge for the future. Energies. 2016;**9**:1-16

[64] J. Leijon, C.B., Freshwater production from the motion of ocean waves – A review. Desalination, 2018. 435: p. 161-171

[65] Babu N, Balaji KK, Nishal S. Wave energy for desalination plants – A Review. International Journal of Engineering Research & Technology (IJERT) ETDM - 2017 Conference Proceedings. 2017;**5**(7)

[66] Hicks DC, Mitcheson GR, Salevan JF. Delbuoy: Ocean wave-powered seawater reverse osmosis desalination system. Desalination. 1989;**73**:81-94

[67] Viola A, Curto D, Franzitta V, Trapanese M. Sea water desalination

and energy consumption: A case study of wave energy converters (WEC) to desalination applications in Sicily. In: OCEANS 2016 MTS/IEEE. Monterey: IEEE; 2016. pp. 1-5

[68] Contestabile PDV. Coastal defence integrating wave-energy-based desalination: A case study in Madagascar. Journal of Marine Science and Engineering, 2018. 6: p. 1-17

[69] Prieto LF, Rodríguez GR, Rodríguez JS. Wave energy to power a desalination plant in the north of Gran Canaria Island: Wave resource, socioeconomic and environmental assessment. Journal of Environmental Management. 2019;**231**:546-551

[70] Schallenberg-Rodríguez J, Del Rio-Gamero B, Melian-Martel N, Alecio TL, Herrera JG. Energy supply of a large size desalination plant using wave energy. Practical case: North of Gran Canaria. Applied Energy. 2020;**278**:1-15

[71] Cabrera P, Folley M, Carta JA. Design and performance simulation comparison of a wave energy-powered and wind-powered modular desalination system. Desalination. 2021;**514**:1-15

[72] Kaczmarczyk M, Tomaszewska B, Bujakowski W. Innovative desalination of geothermal wastewater supported by electricity generated from low-enthalpy geothermal resources. Desalination. 2022;**524**:1-13

[73] Goosen M, Mahmoudi H, Ghaffour N. Water desalination using geothermal energy. Energies. 2010;**3**:1423-1442

[74] Karytsas C, Mendrinos D, Radoglou G. The current geothermal exploration and development of the geothermal field of Milos island in Greece. GHC Bull. 2004;**25**:17-21

[75] Calise F, d'Accadia MD, Macaluso A, Vanoli L, Piacentino A. A novel solar-geothermal trigeneration system integrating water desalination: Design, dynamic simulation and economic assessment. Energy. 2016;**115**:1-15

[76] Noorollahia Y, Taghipoor S, Sajadi B. Geothermal sea water desalination system (GSWDS) using abandonedoil/gas wells. Geothermics. 2017;**67**:66-75

[77] Okati V, Ebrahimi-Moghadam A, Behzadmehr A, Farzaneh-Gord M. Proposal and assessment of a novel hybrid system for water desalination using solar and geothermal energy sources. Desalination. 2019;**467**:229-224

[78] Rashidi MM, Mahariq I, Murshid N, Wongwises S, Mahian O, Nazari MA. Applying wind energy as a clean source for reverse osmosis desalination: A comprehensive review. Alexandria Engineering Journal. 2022;**61**:12977-12989

[79] Fadigas EAFA, Dias JR. Desalination of water by reverse osmosis using gravitational potential energy and wind energy. Desalination. 2009;**237**:140-146

[80] Al-Karaghouli A, Renne D, Kazmerski LL. Solar and wind opportunities for water desalination in the Arab regions. Renewable and Sustainable Energy Reviews. 2009;**13**:2397-2407

[81] Serrano-Tovar T, Suárez BP, Musicki A, de la Fuente Bencomo JA, Cabello V, Giampietro M. Structuring an integrated water-energy-food nexus assessment of a local wind energy desalination system for irrigation. Science of the Total Environment. 2019;**689**:945-957

[82] Rosales-Asensio E, Borge-Diez D, Pérez-Hoyos A, Colmenar-Santos A.

Reduction of water cost for an existing wind-energy-based desalination scheme: A preliminary configuration. Energy. 2019;**167**:548-560

[83] Meli'an-Martel N, del Río-Gamero B, Schallenberg-Rodríguez J. Water cycle driven only by wind energy surplus: Towards 100% renewable energy islands. Desalination. 2021;**515**:1-11

Chapter 7

Improvement of Leucaena (*Leucaena leucocephala*) Benth. Seeds Emergence Using Hot Saline Water Treatment Durations

Yunusa Muhammad Ishiaku, Usman Abdullahi, Haruna Ibrahim, Safiyanu Abdullahi Ahmed and Muhammad Rabiu Hassan

Abstract

Leucaena leucocephala is a leguminous shrub that has the potential for increasing animal production with diverse environmental and ecological significance. An experiment was conducted to investigate the effect of hot saline water treatment durations on the emergence of Leucaena seeds. The experiment was arranged in a completely randomized design with six treatments and three replicates. The treatments are control, soaking of seeds in hot saline water (20 g NaCl/1 L of distilled water) at 80°C for 1, 2, 3, 4 and 5-min durations. The percentage emergence, emergence index, plant height and leave numbers were measured. Hot saline water treatment durations had positive effect of breaking Leucaena seed dormancy and enhance germination. The highest percentage emergence and emergence index (53.9% and 7.95) were obtained at 2 min treatment durations but plant height and number of leaves were highest (6.29 cm and 7.00 cm) respectively in 4 min of treatment durations. It could be concluded that percentage emergence and seedling growth of Leucaena can be enhanced using 2 min hot saline water treatment duration. It is recommended that saline soils that have being degraded due to oil spillage to enable production of Leuceana as animal feed and control environmental degradation.

Keywords: *L. leucocephala*, hot saline water, emergence, treatment durations, ecology

1. Introduction

Rangelands improvement is becoming a thing of the past, despite the fact that 80 to 90% of ruminants depend on the rangeland as their major source of the nutrient [1]. Grazing areas are undergoing extensive degradation due to overgrazing and adverse climatic conditions [2]. Overgrazing of rangelands has led to accelerated soil erosion, reducing productivity and biodiversity of important forages and contributes to the proliferation of non-palatable weeds. To overcome these problems [2], suggested the use of stress-tolerant, soil improver, windbreaker, source of living stakes, high dry matter yielding and nutritive value browse plant as *Leucaena leucocephala*, is an evergreen shrub-like tree that spreads all over the world because of its adaptability to various ecological conditions as it grows in all type of soils from dry, semi-arid and arid areas and tolerate high temperatures for long periods. *Leucaena leucocephala,* to develop sustainable grazing, increase productivity and protect the biodiversity of rangelands. Despite all these advantages associated with Leucaena, the establishment and successful adoption of this multipurpose forage are limited to some extent by water impermeability of the seed coat or 'hardiness', caused by one or more water-impermeable layers of palisade cells in the seed coat [3]. More so, the seed coat can also exert germination restrictive action by its mechanical resistance to radicle protrusion as well as harboring inhibitors to suppress seed germination. Ecologically, this physical dormancy occurring in some plant families of Angiosperms is an important survival strategy as seeds time their germination to coincide with favorable natural conditions to maximize chances of successful establishment of seedlings [3, 4]. This is disadvantageous when a quick, uniform and high germination rate of *L. leucocephala* seed is required, thus discouraging most farmers and grazers from accepting it as a sown pasture species [5]. The presence of a hard seed coat needs to be broken to allow permeability of water and oxygen to reach the embryo and start the germination process. Therefore, numerous techniques to break seed dormancy have been investigated over the years including; soaking seeds in hot water or sulfuric acid or mechanical scarifying/nicking [3, 6–8], as well as varying levels of exposure to dry heat treatments [9]. According to [10], hot water treatment enhances seed germination by affecting various factors, viz., seed coat permeability for gases and water exchange and release of inhibitors. A previous report by [11] indicated that the hot water pretreatment duration to breaking seed dormancy in *L. leucocephala* was 5 min at 80°C. While [7] observed the highest seed germination in Leucaena when the seeds were exposed to hot water at 80°C for 10 min and that soaking in gibberellin 400 ppm is the most efficient in giving the best averages for roots and seedling length. Without scarification [2], reported that only 10% of *L. leucocephala* could germinate. Therefore, the objective of the present study is to investigate the effect of hot saline water treatment durations on the emergence of *L. leucocephala* seeds.

2. Materials and methods

An experiment was conducted at Forage Laboratory of Feeds and Nutrition Research Programme, National Animal Production Research Institute Shika, Ahmadu Bello University, Zaria, Nigeria (latitude 110 12′ N, longitude 070 33′E) and altitude 660 m above sea level in Northern Guinea Savannah of Nigeria [12]. The seeds of *Leucaena leucocephala* were collected from matured plants in the introduction plot. The seeds collected for the experiment were sorted, cleaned and separated from stones and broken pods. The experiment was arranged in a Completely Randomized Design with six treatments and three replicates. The treatment includes T1 (control), soaking of seeds in hot saline water (20 g of NaCl was dissolved in 1 liter of distilled water) at 80°C for 1, 2, 3, 4 and 5-min durations as (T2, T3, T4, T5 and T6 respectively). A total of 450 seeds were sorted, counted and divided into 6 treatments with 75 seeds per treatment and 25 seeds per replicate. The seeds to be treated were wrapped in white clean clothes and placed in 1 L of hot distilled water at 80°C containing 20 g of common salt. The placement was done at the same time and withdrawn at the expiration of the treatment durations and seeds were cooled down under a running tap for 5 min according to [1]. Thereafter, seeds were spread thinly to allow air to dry and were used the next day for germination and emergence test. Twenty-five treated seeds were planted in polyethene bags (15 cm height × 10 cm top and bottom diameters), filled with sandy loam soil to a depth of 7 cm. The polyethene bags were watered daily and seedling emergence was recorded for 15 days. Thereafter, the experiment was terminated and parameters such as percentage emergence, plant height, the number of leaves and emergence index were determined. The emergence index was calculated using [13] formula EI = (TiNi/S), where Ti is the number of days after sowing, Ni is the number of seeds germinated on day Ti and S is the number of seeds sown. Data collected were subjected to analysis of variance and significant means were compared using Duncan's Multiple Range Test of SAS package [14].

3. Results and discussion

Hot saline water treatment durations had a significantly ($P < 0.05$) positive effect of breaking *L. leucocephala* seed dormancy and enhance germination. The highest percentage seedling emergence in **Figure 1** (53.0%) and emergence index in **Figure 2** (7.95) were recorded when *L. leucocephala* seeds are placed in hot saline water at 80°C for 2 min and declined afterwards probably due to an increase in treatment durations. This study shows that *L. leucocephala* seeds are tolerant with regards to salinity and are able to germinate after soaking in hot water containing 20g l^{-1} NaCl for up to 5 min. The study agrees with the report of [15] who listed *L. leucocephala* as a salt tolerant plant. According to [16], salt-tolerant plants are adapted to survive and complete their life cycle under saline levels of higher than 200 mM NaCl. Also, this work agrees with the reports of [17] that the behaviors of *Acacia albida* are tolerant to salinity and are able to germinate after treatment up to 12 g l^{-1} and probably even higher. The Authors also added that the ability to germinate under saline conditions is an important feature for the rehabilitation and reforestation of the species, but is also interesting to use for enhanced soil marginalized and affected by salinization.

The higher percentage emergence reported in this study as compared to 49.0% reported by [2] may be as a result of the addition of salt which may have facilitated

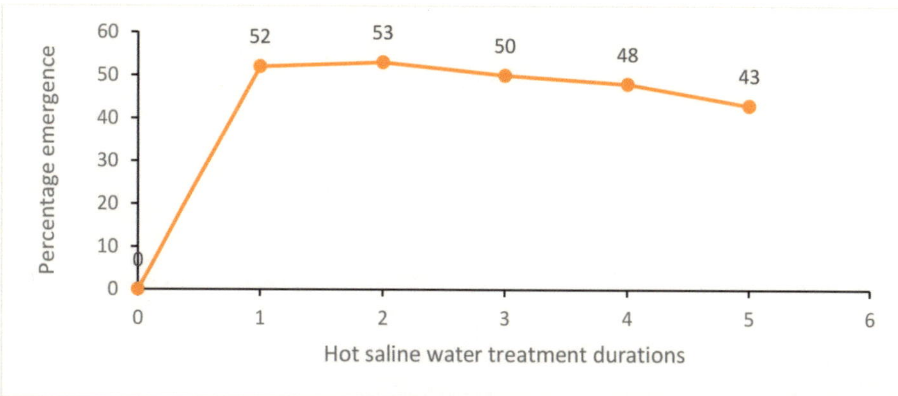

Figure 1.
Percentage emergence of L. leucocephala *seeds as affected by hot saline water treatment durations.*

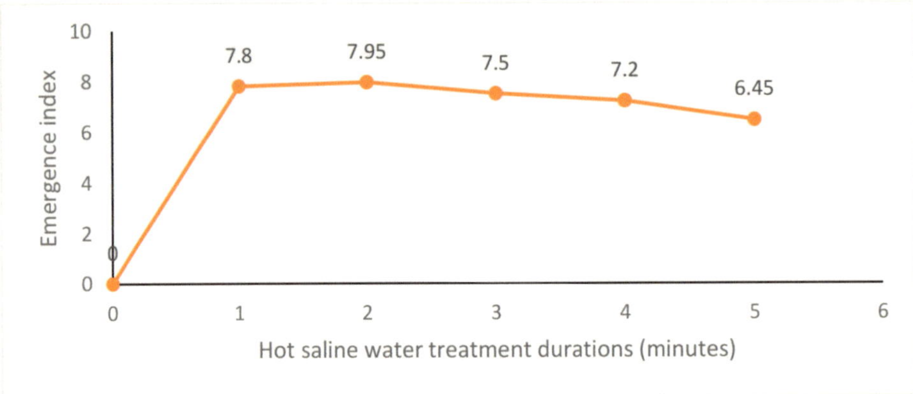

Figure 2.
Emergence index of L. leucocephala *seeds as affected by hot saline water treatment durations.*

the breakdown of hard seed coat and thereby facilitating water permeability and gas exchange which promotes faster germination and seedling emergence. The plant height and number of leaves were significantly ($P < 0.05$) affected by treatment durations. The highest plant height (6.29 cm) in **Figure 3** and the highest number of leaves (7.00 per plant) in **Figure 4** are recorded in 4 min treatment durations. The plant height and the number of leaves reported in this study were within the values 5.30–6.51 cm and higher than 3.63–4.0 leaves per plant earlier reported by [2] for hot water scarifications on seedling emergence and early growth of *L. leucocephala* seeds. The plant height recorded in this study was higher than the ones (2.2 cm) obtained by [7]. Rusdy [2] also observed significant improvement of seed emergence, emergence index and seedling growth of *L. leucocephala* by acid scarification which stimulated prompt and uniform germination compared to hot water treatment durations. Obiazi [5] reported that research has equally shown that hot water treatment can penetrate the seed sufficiently to eradicate bacterial infections inside the seed to promote germination. Also, [10] reported that the treatment of seeds with hot water to improve germination is a better, safe and cost-effective alternative to sulfuric acid and sandpaper methods. Soaking *L. leucocephala* seeds in hot saline water recorded the highest germination rate and the lowest mean germination

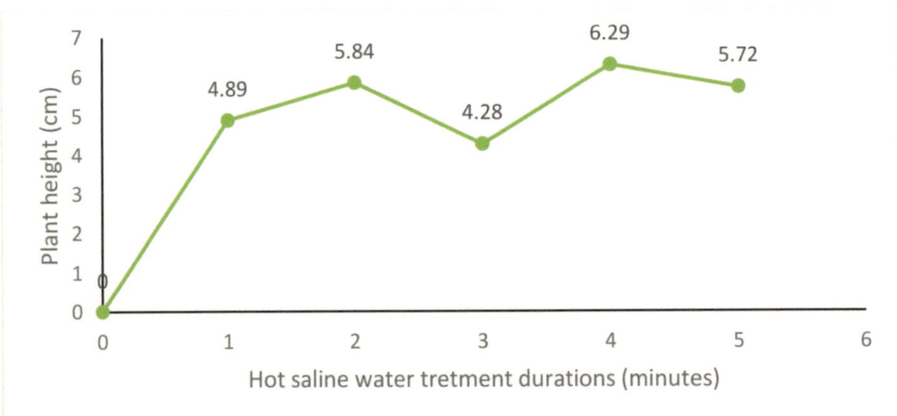

Figure 3.
Plant height of L. leucocephala *seedlings as affected by hot saline water treatment durations.*

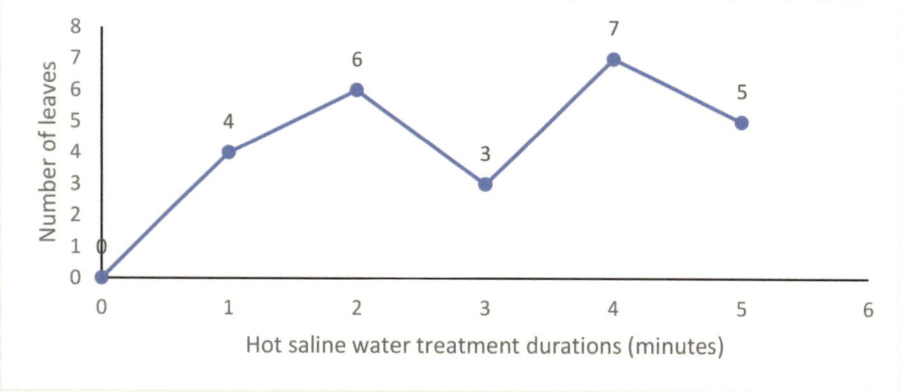

Figure 4.
Leaves number of L. leucocephala *seedlings as affected by hot saline water treatment durations.*

time because it helps in providing the largest area for water absorption and gas exchange for germination. This finding was similar to a study by [11] that soaking of seeds in hot water 80°C for 5, 10 and 15 min gave 94%, 83%, 63% respectively. The variation in germination percentage might be a result of the hardness of the seed casing based on its chemical composition or the environment in which the seeds grow. No germination record was obtained for the control during the 15-day experimental trials. Generally, conversion of agricultural soils into human settlement and industrial use has led to the decreased territory of arable lands. Upcoming climate change with consequences on the raise of sea level, sea water intrusion, and high evaporation was regarded as a major environmental issue which also posed some challenges in the cultivation of economic crops [18] reported that salt stress is one of the serious abiotic factor which limit the growth and development of important crops in agricultural lands. Delvian and Hartanto [19] reported that positive impact of application of Arbuscular mycorrhizal fungi (AMF) towards salt tolerant by *L. leucocephala* with potential application and in salt stressed ecosystem.

4. Conclusion

The concentration of NaCl shows an effect on the germination of *L. leucocephala* seeds. The study indicated that germination and emergence of *Leucaena leucocephala* seeds can be improved by 53.0% when soaked in hot saline water at 80°C for 2 min. Therefore, *L. leucocephala* seeds can be easily propagated in salt stress environments especially in Arid, semi-arid and coastal regions. It is recommended that farmers around these areas are encouraged to adopt this technique but with caution for mass production of this plant for range lands improvement and as dry season feed for livestock.

Acknowledgements

Authors acknowledge the Executive director, National Animal Production Research Institute, Ahmadu Bello University, Shika, Zaria, Nigeria, for his contribution and support to the success this project.

Conflict of interest

The authors declare no conflict of interest from this work.

Author details

Yunusa Muhammad Ishiaku[1]*, Usman Abdullahi[1], Haruna Ibrahim[1], Safiyanu Abdullahi Ahmed[1] and Muhammad Rabiu Hassan[2]

1 National Animal Production Research Institute, Ahmadu Bello University, Shika-Zaria, Nigeria

2 Federal University Gashua, Yobe State, Nigeria

*Address all correspondence to: ishiaku.ym2014@gmail.com; ymishiaku@abu.edu.ng

References

[1] Ishiaku YM, Hassan MR, Tanko RJ, Amodu JT, Abubakar SA, Ahmed SA, et al. Effect of hot and cold-water treatment durations on emergence of Tephrosia (Tephrosia bracteolata Perr. Guil.) seeds. Gashua Journal of Irrigation and Desertification Studies. 2015;**1**(1&2):2489-0030

[2] Rusdy M. Enhancement of seedling emergence and early growth of *Leucaena leucocephala* by hot water, mechanical and acid scarification pre-treatments. International Journal of Applied Environmental Sciences. 2017;**12**(5):857-863

[3] Koobonye M, Maule BV, Mogotsi K. Mechanical scarification and hot water treatments enhance germination of *Leucaena leucocephala* (Lam.) seeds. Livestock Research for Rural Development. 2018;**3**:1. Available from: http://www.lrrd.org/lrrd30/1/kbmo30015.html. [Accessed: March 27, 2021]

[4] Baskin C, Baskin JM. Seeds: Ecology, Biogeography, and Envolution of Dormancy and Germination. San Diego: Academic Press; 2014. pp. 150-162

[5] Obiazi CC. Hot water enhanced germination of *Leucaena leucocephala* seeds in light and dark conditions. Current Research in Agricultural Sciences. 2015;**2**(2):67-72

[6] Tadros MJ, Samarah NH, Alqudah AM. Effect of different pre-sowing seed treatments on the germination of *Leucaena leucocephala* (Lam.) and Acacia farnesiana (L.). New Forests. 2011;**42**(397):1-12

[7] Salih SM, Abdulrrazik AA. Response germination of *Leucaena leucocephala* (Lam.) De Wit Trees Seeds to different treatments. Libyan Journal of Basic Sciences (LJBS). 2020;**10**(1):1-11

[8] Yousif MAI, Wang YR, Dali C. Seed dormancy overcoming and seed coat structure change in *Leucaena leucocephala* and Acacia nilotica. Forest Science and Technology. 2020;**16**(1):18-25. DOI: 10.1080/21580103.2019.1700832

[9] Mott JJ, Cook SJ, Williams RJ. Influence of short duration, high temperature seed treatment on the germination of some tropical and temperate legumes. Tropical Grasslands. 1982;**16**:50-55

[10] Sharma S, Naithani R, Varghese B, Keshavkant S, Naithani SC. Effect of hot water treatment on seed germination of some fast-growing tropical tree. Tropical Forestry. 2008;**24**(3 & 4):49-53

[11] Teles M, Alves A, de Oliveira J, Bezerra A. Procedure for dormancy breakage in *Leucaena leucocephala*. Revista Brasileira de Zootecnia. 2000;**29**(2):387-391

[12] Ovimaps. Ovi location map: Ovi W.A. Review on data analysis methods for seed germination. Crop Science. 2017;**24**:1192-1199

[13] Scott SJ, Jones RA, William WA. Review on data analysis methods for seed germination. Crop Science. 1984;**24**:1192-1199

[14] SAS. Statistical Analysis Software (CD-ROM). Cary, N.C., USA: SAS Institute Inc; 2005

[15] Rashid MM, Hoque AKF, Iftekhar MS. Salt tolerances of some multipurpose trees species as determine

by seed germination. Journal of Biology Sciences. 2004;**4**(3):288-292

[16] Flowers JT, Colmer DT. Salinity tolerance in halophytes. The New Phytologist. 2008;**179**(4):945-963. DOI: 10.1111/j.1469-8137. 02531.x

[17] Karoune S, Kechebar MSA, Belhamra M, Rahmoune C. Study the germination of Acacia albida seeds under salt stress. International Journal of Advanced Research. 2013;**1**(5):276-283

[18] Garcia CL, Dattamudi S, Chanda S, Jayachandra K. Effect of salinity stress and microbial innooculation on glomalin production and plant growth parameters of snap bean (*Phaseolus vulgaris*). Agronomy. 2019;**9**(545):1-21

[19] Delvian D, Hartanto A. Improved salt tolerant of Lamtoro (*Leucaena Leucocephala*) through the application of indigenous Mycorrhiza. Hindawi International Journal of Forestry Research. 2021;**2021**:8100480. DOI: 10.1155/2021/8100480